A PUBLICATION OF THE
NATIONAL RESEARCH COUNCIL OF CANADA MONOGRAPH PUBLISHING PROGRAM

Science, Agriculture, and Food Security

Joseph H. Hulse

Siemens-Hulse International Development Associates Inc.
Ottawa, Canada

NRC · CNRC
NRC RESEARCH PRESS
Ottawa 1995

Publication Proposals: Proposals for the NRC Monograph Publishing Program should be sent to Gerald J. Neville, Head, Monograph Publishing Program, National Research Council of Cnada, NRC Research Press, 1200 Montreal Road, Building M-55, Ottawa, ON K1A 0R6, Canada. Telephone: (613) 993-1513; FAX: (613) 952-7656; E-mail: GERRY.NEVILLE@nrc.ca.

ISBN 0-660-16210-5
NRCC No. 39038

Canadian Cataloguing In Publication Data

Hulse, Joseph H.
Science, agriculture, and food security

Includes an abstract in French. Includes bibliographical references.
ISBN 0-660-16210-5
1. Food supply. 2. Food consumption. 3. Diet. 4. Sustainable agriculture.
I. National Research Council of Canada. II. Title.

HD9000.A1H84 1995 641.3 C95-980212-6

Correct citation for this publication:

Hulse, J.H. 1995. Science, Agriculture, and Food Security. NRC Research Press, Ottawa, Ontario, Canada. 242 p.

Dedication

To Dr. Jacques Diouf:

In grateful recognition of
his exceptional influence on,
and contributions to
food and agricultural research and development,
for the well being of all mankind.

"Agriculture is the only means of making money which can give no offence."

ST. AMBROSE OF MILAN, *De Officiis*, 374 AD

"Let man consider the sources of his food, that we pour down waters and cleave the earth...thereupon we cause grain to grow...for you and for your animals to enjoy."

KORAN 80, 24-32

"And thou shalt be secure, because there is hope"

JOB xi, 18

Table of Contents

Abstract/Résumé

Food security connotes consistent access to food sufficient for bodily health and activities. Food security should be counted as a basic human right.

Conceptions of food security are not enlightened by resort to global agronomic arithmetic: how many organisms might be sustained by hypothetical estimates of the planet's productivity. Food security is best practically arrived at by shrewd assessment of the needs and resources of families and discrete communities.

The author discusses differing concepts of what diets are perceived as adequate or inadequate, well- or ill-balanced, for various age and sex categories and levels of activity. Philosophies and prescriptions on diet and health posited by the U.S. Surgeon-General, and a committee of the National Academy of Sciences are examined.

Food security need not be congruent with self-sufficiency: the capacity of a nation or community to satisfy total nutritional need from the resources within its territorial boundaries. Food security, communally and individually is demonstrably dependent on comparative prosperity, power and poverty. Among resource-deficient nations, persons with wealth or political power rarely suffer chronic hunger.

The text deals with technical, economic, social and political issues critical to food security; conditions that may provoke or aggravate insecurity. It discusses diets prescribed for unusual environmental and work situations drawing on physiological studies of "captive" military subjects and astronauts. It describes unconventional and eccentric dietary prescriptions; various food sensitivities and intolerances.

The author argues for more systematic analysis of the complexities of food security, with less over-reaching reliance on technologies of production, and broader consideration of all that is entailed between farms or fish-landings and consumers. Urban populations, now rapidly expanding throughout the world, cannot be sustained by primitive systems of production, unsupported by efficient means of preservation and distribution.

The author discusses divergent concepts of "sustainable agriculture", proposed not infrequently to poor countries by advocates whose own security will be immune to the risks inevitably imposed. He examines resources and policies that influence economically efficient and ecologically conservative food production and distribution. The author discusses soil and water management, genetic diversity, atmosphere and climate, various production systems as these relate to food security.

The author points to insufficient concern for and investment by donor and development agencies in post-production systems: in studies of genetic diversity relevant to utility; post-harvest preservation, transformation, transportation, distribution and marketing; seriously inadequate infrastructures and essential services between rural and urban communities. Assurance of consistently reliable food quality and safety assumes greater complexity as genetic modifications to crops and livestock create unfamiliar food materials; as spatial and temporal distances between producers and consumers progressively expand.

The author reviews proposed research priorities for agricultural, food and nutritional sciences. Finally, the text characterises requisite transformations in patterns of human resource development: a compelling need for food systems analysts and planners - men and women trained to integrate systematically the immensely diverse complexity of conditions and components contingent upon food security.

La notion de sécurité alimentaire comporte un accès continu à des quantités de nourriture suffisantes à la santé et à l'activité corporelles. La sécurité alimentaire devrait compter parmi les droits fondamentaux de la personne.

Les mesures de la sécurité alimentaire ne sont pas éclairées par le recours à une arithmétique agronomique globale : combien d'organismes peuvent être nourris par des estimations hypothétiques de la productivité de l'ensemble de la planète. On parvient à des mesures plus exactes en évaluant judicieusement les besoins et les ressources des familles et des collectivités.

L'auteur aborde des notions divergentes au sujet de ce qu'est un régime alimentaire adéquat et inadéquat, bien ou mal équilibré pour les divers groupes d'âges, sexes et niveaux d'activité. Il passe en revue les préceptes et prescriptions énoncés par le Surgeon-General des États-Unis et un comité de l'Académie nationale des sciences en matière d'alimentation et de santé.

La sécurité alimentaire ne signifie pas nécessairement l'auto-suffisance alimentaire, c'est-à-dire la capacité d'une nation ou d'une collectivité à satisfaire à ses besoins totaux d'alimentation à même les ressources produites sur son territoire. Comme on peut le démontrer, la sécurité alimentaire, tant communautaire qu'individuelle, dépend de la prospérité, du pouvoir et de la pauvreté. Dans les nations pauvres en ressources naturelles, les personnes économiquement ou politiquement bien nanties souffrent rarement de la faim chronique.

Le texte traite des problèmes techniques, économiques, sociaux et politiques au coeur de la sécurité alimentaire et des conditions qui peuvent provoquer ou aggraver l'insécurité. Il expose les régimes prescrits en milieu inusité et pour des conditions de travail inhabituelles, puisant à même les études physiologiques effectuées auprès de militaires et d'astronautes «captifs». Il décrit des prescriptions non conventionnelles et excentriques ainsi que diverses sensibilités et intolérances à certains aliments.

L'auteur préconise une analyse plus systématique des ramifications de la notion de sécurité alimentaire qui s'appuierait moins sur les technologies de production et qui tiendrait davantage compte de tout ce qui intervient entre les fermes ou les ports de débarquement du poisson et les consommateurs. Les populations urbaines, maintenant en pleine expansion partout dans le monde, ne peuvent pas être approvisionnées par des systèmes primitifs de production non appuyés par des moyens efficaces de conservation et de distribution.

L'auteur examine diverses notions d'«agriculture soutenable», proposées assez fréquemment aux pays pauvres par des promoteurs dont la propre sécurité sera bien à l'abri des risques inévitables que ces mesures entraînent. Il étudie les ressources et les politiques qui influent sur la production et la distribution des aliments selon des méthodes économiquement efficaces et écologiquement rentables. L'auteur aborde la gestion des sols et des eaux, la diversité génétique, l'atmosphère et le climat, les divers systèmes de production dans leurs relations avec la sécurité alimentaire.

L'auteur fait ressortir l'indifférence des donneurs et des agences de développement à l'égard des systèmes post-production et le manque d'investissements à ce chapitre : études sur la diversité génétique pertinente à l'utilité; projets de conservation, de transformation, de transport, de distribution et de mise en marché des récoltes; mesures pour pallier aux graves lacunes des infrastructures et des services essentiels entre les collectivités rurales et urbaines. Il devient de plus en plus difficile de garantir la qualité et la sécurité des aliments au fur et à mesure que les modifications génétiques aux récoltes et aux animaux d'élevage produisent des denrées inconnues et que les distances spatio-temporelles entre producteurs et moissonneurs s'accroissent.

L'auteur fait une revue des priorités proposées dans le domaine des sciences de l'agriculture, de l'alimentation et de la nutrition. Enfin, le texte dresse les caractéristiques des transformations requises dans les modèles de développement des ressources humaines : un besoin pressant d'analystes et de planificateurs des systèmes d'alimentation, des hommes et des femmes formés en vue de procéder à une intégration systématique des conditions et des composantes immensément complexes de la problématique de la sécurité alimentaire.

Acknowledgements

It was the late Professor John Kinsella who, shortly before his sad and premature death, persuaded me to write this book. Professor Kinsella was a scientist of outstanding capability, a gentleman of exceptional generosity of spirit.

I wish, gratefully, to acknowledge the many information officers, librarians and others, in the following organisations who kindly provided access to their literary and computerised collections, to many publications, together with much helpful advice and assistance: Agriculture Canada; the Canadian International Development Agency; the Canadian Medical Association; the Canadian National Institute of Nutrition; Health and Welfare Canada; the Canadian Department of National Defence; the International Development Research Centre; FAO in Rome; the International Council of Scientific Unions in Paris; UNDP in New York; UNICEF in New Delhi; the M S Swaminathan Research Foundation; the Indian Agricultural Research Institute;the World Bank; the Australian Academy of Technological and Engineering Sciences; the Natural Resources Institute in Britain; the Rockefeller Foundation; the United States Embassy in Ottawa; the US Department of Agriculture; the US Food and Drug Administration; the US National Aeronautic and Space Administration; USAID and the USAID CRSP Program; the Winrock Foundation; the World Resources Institute. Particularly helpful have been the many scientists and information officers in the CGIAR International Agricultural Research Centres, the CGIAR and Technical Advisory Committee Secretariats, who provided an immense volume of annual reports and other publications of relevance to the subjects discussed.

Dr. Alexander F. McCalla's willingness to write the Foreword is especially appreciated.

As in all such publications, there are practical limits to scope and dimension. Given the overwhelming volume of literature devoted to food security, its related sciences, technologies, social and economic issues, it was, throughout, as difficult to decide what to omit as what to include. To the many institutions and authors who, justifiably, will consider that their thoughtful and relevant publications have been overlooked or inadequately recorded, I express unreserved regrets.

My very sincere thanks go to the staff at NRC Research Press of the National Research Council of Canada, for helpful editorial advice and guidance, and for the imaginative design of the front cover: a collage composed of coloured slides selected from the author's personal collection.

To all those who have provided support, assistance and encouragement, most particularly to my wife Adeline, I am indeed deeply grateful.

JOSEPH H. HULSE
August 1995

Foreword

The issue of whether the world has the capacity to feed an ever-growing population has been at the forefront of the global debate since time immemorial. The current form of the debate, as a race between food production and population growth, stems from the writing of Robert Thomas Malthus nearly 200 years ago. The Malthusian crisis was held off for the first hundred years by expanding the area of production. In the last 100 years it has been steady, and sometimes spectacular, contributions from agricultural science that have carried the day.

With the world's population likely to reach in excess of 8 billion people in the next 30 years, the topic addressed by Dr. Hulse is timely and important. The treatment of the subject is comprehensive and pointed. Those who know Joe Hulse know that he says what he means precisely, with conviction and with passion. This book has these characteristics which make it challenging and readable.

Dr. Hulse's lifetime experiences, first in the food industry and then in international agricultural development, show through clearly. For example, the primacy given to food consumption and adequate diets clearly show his passion for the food security of the individual as a basic human right. His extensive treatment of post production systems is consistent with his lifelong crusade for greater research attention to this topic.

His skepticism about global projections clearly flows from his mistrust of long-term economic forecasts and his strong belief in the ingenuity of farmers to produce efficiently, if given proper incentives and access to appropriate technology. Thus, his extensive discussion of sustainable agriculture and research priorities is welcome and timely.

There is much more I could say but that would delay the reading of the book that follows. It is vintage Joe Hulse, filled with facts, hypotheses, assertions and controversies. It is a welcome addition to a topic which in my view will become more critical in the 21st century. The sometimes complacent, short-term view currently held by many, needs to be challenged. This book goes a considerable way in widening and deepening the debate.

ALEXANDER F. MCCALLA
Director, Agriculture and Natural Resources Department
The World Bank, Washington, DC.

OVERVIEW AND COMMENTARY

To ensure food security for all people is the declared purpose of the Food and Agriculture Organization of the United Nations (FAO), and a high priority of various other agencies. This book examines the differing concepts of food security and the practicalities, policies and resources that shape issues of food security.

Many definitions relate food security to access to an adequate diet, but the authorities differ significantly on what constitutes an adequate diet: which specific nutrients, known or believed to be essential, are required and at what levels of intake. Recommended levels of intake vary among governments and international agencies, and within particular agencies over time. Some prescribed levels are stated with a precious degree of accuracy. It is not clear, for example, why a Canadian adult male aged 51 needs one less gram of protein each day than a similar man aged 49. It is extremely unlikely that such a difference could be determined or monitored under normal living conditions.

Diets that are perceived as satisfactorily adequate, as unsuitable, or inadequate are discussed relative to criteria of good health. The text discusses those who have access to too little food, those who eat too much, and those whose diets appear nutritionally ill-balanced. Access to too little is associated with poverty; however, the definitions and precise quantifications of poverty in relation to nutritional adequacy vary substantially. Inequalities and inequities in income, within and among nations and communities, have a dismal history and there is evidence that the economic gap between the rich and the poor is widening, globally and even within some affluent nations (Economist 1994).

In the author's opinion, efforts to predict global carrying capacity (how many people the planet can adequately feed) serve very little practical purpose, as illustrated by the conflicting and widely divergent hypotheses. Farmers who husband beef or dairy cattle can determine how many animals can be sustained on any particular sward of pasture grasses and legumes; but humans are not grazing animals, and how many people the planet can sustain is indeterminate and hypothetical. It is for specific communities and families that resources essential to food security are best assessed, prescribed and provided.

Food security need not be synonymous with self-sufficiency, as is illustrated in the extreme by Hong Kong and Singapore. There is much controversy concerning the degree to which developing nations will be able to satisfy their nutritional needs from indigenous resources, and the extent to which they must depend on food imports from more prolific producers (Brown and Kane 1993; Carruthers 1993; Mitchell and Ingco 1993). The optimists and pessimists differ because of widely variant assumptions. (Thomas Malthus proposed that everyone should enjoy at least one meal each day that included roast meat and wine.) On the basis of present patterns of consumption, the planet's resources could feed many more Kalahari bush people than Californians or Canadians.

Forecasts of carrying capacities derive largely from estimates of primary production (by what amount grain yields might be increased) against predicted population growth. Few appear explicitly to take account of the complex constraints imposed by the seemingly irreversible expansion of city populations in developing regions, or of changing demand consequent upon the rising disposable incomes evident in Asia, the Near East, Latin America and several African countries. Donor and development agencies pay little attention to the resources needed to deliver food safely from rural producers to urban markets.

The World Bank's Development Report for 1994 paints a grim picture of the insufficiencies and deterioration among infrastructures essential to rural transport and communications. There are serious inadequacies in many industrial processing, distribution and marketing facilities, and in the energy, transport and infrastructural amenities and services needed to carry essential food from farmers to consumers; safely and economically from regions and seasons of abundance to those of scarcity. Densely populated urban communities cannot be assured of food security from subsistence-scale farming systems or from primitive domestic and cottage processing industries.

Military personnel employed under strict discipline have proven convenient experimental subjects in studies of physiological response to extreme physical, environmental and dietary conditions. Case studies of severe nutrient and water deprivations are described. Accounts of various intolerances, allergic responses and aversions to foods are also discussed, as well as unusual diets, many originating in the USA, and seemingly offered by their proponents as today's equivalents of the alchemists' *Elixir vitae*.

Diverse concepts of "sustainable agriculture" are drawn from a vast literature exemplifying the tensions that exist in severely stressed ecologies between the need for sustenance and the pressure to conserve the environment. It is not helpful to equate "sustainable" to any specific system of production, or to propose any system as a global panacea. Sustainable agriculture requires research to arrive at collectively reinforcing technologies, services and public policies. What is sustainable depends on local demands, traditions, resources, opportunities and constraints. What must above all be sustained are the healthy active lives of people. Among severely stressed ecologies, the livelihoods and sustenance of poor people may often be in conflict with sustainment of natural resources. In absence of fuels they can afford, rural folk will chop down trees.

Public policies markedly influence opportunities for food security. Policies that legislate subsidisation of production and exports for ephemeral local political advantage have caused severe economic and ecological degradation among less fortunate nations. Various policies that purport to conduce to food security for the poor are reviewed.

Resources and conditions that influence food security are examined, including atmosphere and climate, land and water, energy in production, processing and distribution, and pest and disease management. Biodiversity is discussed relative to natural and managed resources and the transgenic modification of food crops, aquatic and terrestrial animals. People who ideologically oppose all induced transgenesis might be reminded that genetic transformations realised in a laboratory reflect genetic evolutions that have come about naturally over countless millenia.

The Consultative Group on International Agricultural Research (CGIAR) family of international agricultural research centres continue to make unique contributions to food security by raising the yield potential of the principal food crops, by diagnosing and designing conservative, productive farming systems, and by enhancing the capability of national agricultural institutions (Collinson and Tollens 1994). For instance, adapting genotypes and agronomic systems developed at the International Rice Research Institute (IRRI), farmers in 11 Asian countries increased their average rice yields from 2 to 3 t/ha between 1970 and 1990. Similar on-farm increases are reported for other food crops improved by the family of international research centres.

Benefits from agricultural research derive from a slow process of infiltration and diffusion, not by "instant impact". Crop improvement research can never end, since species must be progressively modified to resist new pests and mutations, and to adapt to differing ecologies. Donors who financially support agricultural research must be prepared for protracted, persistent investment over many decades.

Priorities for the future should include: research to raise sustainable yields and productivity, and research related to post-harvest systems. Germ plasm collections held by international agricultural research centres have accessioned over 600,000 identified genotypes of the principal food crops. Although agronomic and phenotypic characters are well documented, there is comparatively little information about inheritable useful functional properties which are of critical importance to processors. To provide the consistent quality demanded by consumers, processors must be provided with biological materials that are consistent in essential functional properties. Genotypes with superior agronomic characters but that lack desirable functional properties will likely remain little more than biological curiosities. Biological diversity underutilised is diversity wasted. There is a clear need for more effective consultation and cooperation between plant breeders and agronomists on the one side and food scientists and technologists on the other.

Productivity must increase faster than population growth, but higher crop and livestock yields alone will not satisfy future demand. Urban food security requires economic, technologically efficient food processing and distributing industries to provide foods that are safe, stable, convenient and acceptable to urban consumers. Small-scale industries may satisfy the needs of rural people who live in close proximity. The greater the distance in time and space between producers and consumers, the greater the need for efficient industries of a size that can provide foods sufficient in quantity and diversity, foods reliably consistent in nutritional quality and wholesomeness. It is difficult enough to prevent food-borne infections and intoxications among technologically advanced countries in temperate climates; to provide safe foods to city dwellers in tropical ecologies presents a formidable challenge to both governments and food industries.

Inconsistent, inadequate and disorderly integration of production with post-production systems persists as a primary constraint to food security in many developing countries. Post-production systems that embody all that obtains between rural farms and urban consumers appear as the Cinderella of food security planning. In common with farming enterprises, industries that process, market and distribute foods need supportive research, technical advice, relevantly trained staff, stimulative policies and incentives. As well as providing consistently safe foods, industries that process and distribute food and perform related services employ more people than any other industry in the USA, Canada, India, Malaysia, Thailand and many other countries.

Food marketing services receive little government support except where they impinge on international commodity trade. Transport systems and their essential infrastructures are deteriorating in most developing regions and in several more affluent nations (World Bank 1994).

Sustainable food security is destined for the welfare of people, and is brought to reality by people, yet many publications and development agencies focus more upon technologies, policies and ideologies than upon the people to be sustained and the people needed to ensure their sustenance. "Sustainable" is an overused adjective in development literature. The verb, "to sustain" derived from the Latin "sustenire", is an active, transitive verb: someone or some agency must act persistently to sustain whatever is to be sustained. Essential courses of action will vary among locations and within locations as available resources, prevailing agro-ecologies, social and economic conditions differ and alter. Sustainable agriculture knows no global panacea.

Few donor or development agencies provide sustained support to any project long enough for full realisation. Development, however defined, is a slow process, requiring patient, persistent analysis, support and review. The reasonable recommendations of the Pearson, Brandt and Brundtland Commissions, while enthusiastically acclaimed at the time of promulgation, were soon forgotten or ignored.

Holistic, interdisciplinary, integrated management is essential to economic, social, technological and industrial development. A salutary example of interactive integration is found among organisations now studying the influence of complex global cycles on climate change. The pursuit of food security deserves no less extensive and systematic an integration of essential skills and disciplines.

Food security for the poorest nations calls for intergovernmental and inter-agency cooperation of an intensity manifested in war time. One international agency in particular has the ability to provide the necessary leadership and coordinating mechanisms. FAO, which celebrates its 50th anniversary in 1995, possesses unique access to current information on the state of food and agriculture around the world. From its vast bank of information, fed by its extensive network of regional and country offices, FAO can identify changing circumstances, demands and resources as they impinge upon present and future food security. FAO can predict where climatic, political and other disturbances may bear upon local and regional food security.

To fulfil the role of world leader, FAO should focus its energies on defining food security and agricultural policies for both national governments and regional cooperative alliances. Much of the specific technical assistance needed and identified by FAO could be economically provided by other agencies.

To formulate these essential policies, FAO should seek to cooperate more closely and integrate its activities with competent bilateral, voluntary and nongovernmental agencies, with the World Bank, with IFPRI, and with private industrial associations. Policies could then be defined for agro-industrial development and for efficient systems of food preservation, processing and distribution, taking account of essential supporting infrastructures and services. FAO's member governments must accept that FAO cannot be all things to all nations, that it must concentrate on activities for which it is most favourably endowed and equipped.

Finally, overly optimistic prognostications and predictions relative to food security for the world's poorest nations and communities should be treated with caution. Though remarkable progress in agricultural improvement has been made and, with sustained adequate support, will continue to be made, sustainable food security for all people is far from assured.

Affluent nations have devised and instituted enormously expensive alliances for their collective military security. How much more benign, humane and civilised it would be for these same nations unselfishly to cooperate in an alliance to assure sustainable food security for all mankind.

CHAPTER 1. FOOD SECURITY

INTRODUCTION

Food security is more easily discussed in general than in specific terms. In principle, a state of food security is one in which the individuals or communities under consideration enjoy consistent access to foods that in quantity, quality and composition provide a hygienic and nutritionally adequate diet. This publication starts with definitions and concepts of critical descriptive terms. It examines concepts of what constitute adequate diets, taking account of extremes of excess, insufficiency, and imbalance. It goes on to describe and discuss the various constituent components of food and agricultural production and post-production that must be integrated into efficient and economic systems of supply, preservation, distribution and utilisation.

Food security, however conceived, is complex and controversial. While certain principles are valid, for each community and indeed for each individual family, the specific needs and the resources to satisfy those needs will differ. Definitions of food security therefore tend more to generalisations than to specific prescriptions. As is discussed subsequently, food security cannot be assured by science and technology alone. Systematic scientific research elucidates essential basic principles on which to base technologies to develop and diversify stable systems of food production, distribution and utilisation. To function for human benefit, science and technology need a congenial political, social and economic environment. Therefore, this book addresses policy issues as well as science and technology.

Food security is not synonymous with self-sufficiency. Singapore and Hong Kong, which produce very little food by crop cultivation or animal husbandry, earn enough foreign income from the goods and services they sell to buy from other nations the food their citizens need.

Definitions and Concepts

Lord Kelvin, who with Joule unravelled the mysteries of evaporative cooling, once remarked: "If you can define precisely and measure accurately that of which you speak, and express it by a number, you know something of your subject. If you cannot accurately define and measure, your knowledge is meagre and unsatisfactory."

Not everyone who writes about nutrition and related subjects is precise in terminological usage or in defining the terms they use. Articles written for a general non-specialist readership often fail to distinguish between acute, chronic and transitory hunger. "Famine", "starvation", "malnutrition", "hunger" and other descriptors of nutritional deficiencies tend to be used interchangeably and indiscriminately.

"Security" derives from the Latin *securus*: "free from care". People who enjoy food security are free of the risks of chronic hunger, malnutrition, starvation or famine. Since these four nouns tend to be used synonymously it may be helpful to define each of them.

Famine describes an extreme general scarcity of food among large populations. *Starvation*, derived from an Old English word, originally meant to die a slow lingering death, from cold, disease, from lack of food and/or water. *Malnutrition* defines an inadequacy or imbalance in intake of essential nutrients. (Latin *nutrire*: to nurse, suckle, feed). *Hunger* describes the painful or uncomfortable sensation caused by want of food. It covers a transitory condition experienced by all who enjoy healthy appetites; and varying degrees of chronic malaise caused by serious lack of food. FAO's imaginative Freedom From Hunger Campaign was intended to alleviate famine, starvation and extreme chronic malnutrition, not the transitory hunger experienced by active healthy men and women.

Starvation, Malnutrition and Food Security

Stevenson (1958) defined acute starvation as a complete lack of food of caloric value; chronic, partial or semi-starvation as a state in which some calories are available, though quantitatively less than are required for health and activity. Malnutrition he defined as a partial deficiency of one or several essential nutrients such as protein, amino acids, fatty acids, vitamins and minerals.

The Select Committee on Hunger (1990) defined Food Security as: "All people obtaining a culturally acceptable, nutritionally adequate diet, through non-emergency food sources at all times".

During the International Conference on Nutrition convened by FAO and WHO in December 1992, the "Plan of Action" contained the following statement: "Food security is defined in its most basic form as access by all people at all times to the food needed for a healthy life. Achieving food security has three dimensions. First, it is necessary to ensure a safe and nutritionally adequate food supply both at national and household levels. Second, a reasonable degree of stability in the supply of food throughout each year and from one year to the next. Third, and most critical, each household must have physical, social and economic access to enough food to meet its needs. This means that each household has the knowledge and the means to produce or procure the food that it needs on a sustainable basis." The FAO/WHO statement emphasises the need for fair distribution of food among communities, among households, and within each household so that each member is adequately fed. It insists that food security be both a fundamental objective and a measure of the success of national and international development policies (FAO/WHO 1992). Access to a safe and adequate diet has long been considered a basic human right.

The Brundtland Commission Panel on Agriculture, Food, Forestry and Environment speaks of Sustainable Livelihood Security (WCED 1987). Livelihood is defined as adequate stocks and flows of food and cash to meet basic needs. Security refers to secure ownership of, or access to, resources and income-earning activities, including reserves and assets, to offset risk, ease shocks and meet contingencies. Sustainable refers to the long-term maintenance or enhancement of resource productivity. Families and communities may realise SLS in various ways: through self-sufficient subsistence farming; through ownership of livestock, land and/or trees, products of which have a market value; from fishing; through income gained from regular full-time or part-time employment. The Panel on Food Security of the Brundtland Commission perceived "Food Security" as all people having access to sufficient stocks and supplies of food to provide a nutritionally adequate diet. This need not entail a state of total self-sufficiency. It does not require that everyone become a subsistence farmer. It does require that all who are not farmers have money sufficient or goods they can barter for essential foods.

Food and Nutrition Policies

During the mid-1930's, the League of Nations Committee on Nutrition made the following observation: "The movement towards better nutrition in the past has been largely the result of unconscious and instinctive groping of men for a better and more abundant life. What is now needed is the conscious direction towards better nutrition. Such direction constitutes policy. Nutrition policy must be directed towards two mutually dependent aims: first, *consumption*, bringing essential foods within reach of all sections of the world community; second, *supply*." (League of Nations 1936, 1937).

The Committee's report goes on to urge (a) recognition of nutrition policy as of primary national importance; (b) better education on human nutrition; (c) a more equitable distribution of income since it is the poorest people who are the most nutritionally deprived.

A report by the U.S. NRC of a study of world nutrition is in broad agreement with the League of Nations (NRC/NAS 1977). It states: "An important cause of malnutrition is absence of policies and programs that foster the best use of available food supplies. In poor and rich countries alike, governments continually make decisions that affect nutritional status with little knowledge of nutritional consequences. Success in alleviating hunger and malnutrition will depend upon increasing the supply of the right kinds of foods, reducing poverty, improving the stability of food supplies, and decreasing the rate of population growth."

Reutlinger (1987) defined food security as "Access by all people at all times to enough food for an active, healthy life". He estimated that some 730 million people, most in Asia and sub-Saharan Africa, suffer food insecurity because of diets inadequate to sustain active working lives. Reutlinger suggests that food insecurity results from improper macro-economic policies, and local economic and political structures that restrict the ability of households and individuals to purchase food sufficient for need. Reutlinger points out that food insecurity may be transitory or chronic, a chronic condition being of long duration, and a consequence of inadequate resources with which to produce or purchase food. Transitory hunger is temporary, resulting from unstable or disrupted food production, a sharp rise in food prices or a drop in household disposable income.

In an extreme extended form, transitory hunger can exacerbate into famine. Such has been the case in drought stricken countries deprived of access to strategic stocks or external food aid in time sufficient to prevent catastrophic malnutrition. Famines can ravage when global food stocks are plentiful. The Bengal famine in 1943 was aggravated by a rapid rise in urban incomes that lifted the price of a limited food supply to heights poor rural people could not afford.

Sen (1987) related food security to entitlements and ownerships and contended that consideration of food security must focus upon adequate access at household and individual levels rather than estimates of national aggregate demand. Sen addressed entitlements as they are accepted in market economies, including: voluntary trade; production in terms of entitlement to own what they as individuals or communities have produced with their own resources; entitlement in relation to labour and inheritance. Food security entitlement depends on fair exchange and trade of goods and services among those who are gainfully employed. For the poor and unemployed, food security relies in some degree on social programs and provisions. Sen argued that chronic malnutrition and hunger are less a function of total food supply than of how equitably the supply is distributed. It is difficult for poor, chronically hungry and malnourished folk to establish entitlement to an adequate food intake.

The extent and effects of food insecurity may differ if considered simply in terms of the calories required to stay alive, or of diets sufficient to sustain work and healthy activity. Macroeconomic policies must take account of incomes sufficient to satisfy citizens' needs, and measures to stimulate sustained economic growth to allow for expanding and diversifying demands. Food policies must be integral to each nation's overall economic policies.

Food Security in the USA: The Critical Dimensions

Busch and Lacy (1984) in an extensive critical review of agriculture and food production in the USA conceived food security as having at least three dimensions: Availability, Accessibility, Adequacy.

1) Availability requires that there be a stable sustainable system of production and distribution to provide food sufficient to satisfy the needs of all dependent people, and resilient enough to survive the natural climatic and human disruptions

and disturbances by which crop and livestock production can be adversely affected.

2) Accessibility must not be limited by what economists describe as "effective demand". Populations of all income and social categories, rural and urban, must have access to a nutritionally adequate supply of food. In a free market economy, all must have sufficient disposable income to afford the purchase of food sufficient for their needs.

3) Adequacy can be described in general terms as a food supply that satisfies different nutritional needs among various conditions of men, women and infants, young and old, rich and poor, who make up national and communal populations.

Consistent with the stated intention to define, and where possible, to quantify terms used it may be helpful to examine the roots, origins and dictionary definitions of the descriptors used by Busch and Lacy.

The root of "Availability" lies in the Latin *valere*: translated as to avail, to be of value, to be effective for a purpose. "Available" means being capable of producing a desired result; "availability" that which is at disposal to produce a desired result. "Accessible" has its root in the Latin *accedere* meaning to arrive at; capable of being reached or used. "Adequate" derives from the Latin *adaequare*, defined as "towards equality"; "equal in extent".

Busch and Lacy's definitions are practically useful. "Availability" is interpreted to mean that, overall, food is produced in quantity sufficient, if equitably distributed, to produce the desired result of satisfying all nutritional needs. "Accessibility" means that food sufficient to satisfy is within the reach of all who have need. In an economic system wherein all food must be paid for at market prices, the very poor may not have access though sufficient food is available. "Adequate" in the Busch/Lacy sense refers to nutritional adequacy; the food available and accessible is nutritionally equal to collective and individual physiological need.

The Human Development Index

UNDP's 1993 Human Development Report states that security should be reinterpreted as security for people, not security for nations or land. It goes on to say that development must be woven around people, the purpose of development being to enlarge peoples choices (UNDP 1993). The Report again revises the criteria and parameters by which the Human Development Index (HDI) is defined. HDI is an index according to which all nations are classified. HDI reflects the state of effective purchasing power, education, and health, and offers a more comprehensive integrated indicator than GDP or GNP, particularly when these latter are expressed in United States dollars at official rates of exchange.

Each nation's assessed HDI embodies three important indicators of human well being: Longevity, Knowledge, and Income, each of which, in various ways impinges upon individual access to resources essential to food security. Longevity is looked upon as life expectancy at birth; Knowledge is equated with adult literacy and mean years of schooling; Income is derived from a complex formula that relates income to the accepted poverty level. For each nation, the HDI is adjusted according to income distribution.

The Report reemphasises that poverty, exacerbated in many countries by spreading unemployment, is the primary cause of food insecurity. To encourage economic growth along with employment opportunities, the Report suggests that governments invest in basic education and training of workers; stimulate private enterprise; make markets accessible to everyone; and support small-scale enterprises and informal employment through provision of credit and fiscal incentives. Employment safety nets may be extended by labour-intensive public works programs at times of major economic distress.

The Report is critical of the way international aid, particularly bilateral official development assistance, is allocated, too much being invested in support of political and strategic objectives, too little for assistance to the world's poor. To alleviate poverty and bring food security to a greater number, aid should be directed at such basic issues as health, education, environmental conservation, and population control.

While recognising the progress made in recent years in raising life expectancy, access to health services, increased food production per capita on all continents but Africa, rapid economic growth in south and east Asia, reduced infant mortality; the Report points to many millions of elderly without income security; to the 800 million chronically undernourished; to 12 million children who die annually from diseases related to malnutrition; to the high global incidence of illiteracy; and to the 1.3 billion who live below defined poverty lines.

When income distribution is considered, several seemingly prosperous nations are found to include sizable numbers whose standard of living is equivalent to that of people in the Third World nations. For example, the white population of the USA, when disaggregated from the total, enjoy the highest HDI in the world. In contrast, blacks and Hispanics in the USA rank below number 30 in the HDI listings, their index being equivalent to that of Trinidad and Tobago. In India, Kerala and the Punjab are rated much higher than such poor states as Bihar and Uttar Pradesh. Brazil, Panama and Mexico display wide disparities between rich and poor citizens. Consequently a single index of a nation's level of development or food security may not reveal vast differences in status among inhabitants. Both within and among nations there is not equitable access to food security, nor to health services, education, knowledge, technology or information, all of which bear upon individual prosperity or poverty, and access to a nutritionally sound diet (UNDP 1993).

Nutritional Status

Nutritional status has been defined as a person's state of health as it is influenced by intake and utilisation of nutrients. An optimal and acceptable nutritional status should be realised if sufficient but not excessive sources of energy and essential nutrients are consumed, and antinutrients, toxins and other deleterious substances are avoided.

Until recently, nutritional studies concentrated on specific nutrient deficiencies and such diagnosable diseases as scurvy and pellagra, caused by too little ascorbic acid and niacin, respectively. Specific deficiencies result from low intakes or inefficient utilisation of one or more essential nutrients. Unsatisfactory nutritional status can also result from over consumption of particular nutrients or from eating too much while taking too little physical exercise.

Among populations and individuals moving freely as they live their lives and go about their activities, the difficulties of determining nutritional status are many and complex. Clinical determinations of precise nutrient intake, utilisation and balance require that subjects under study be confined to a metabolic ward or some other controlled environment. Consequently, many reported data of nutritional status rely upon interviews or questionnaires in which large sample populations are asked to recall and record the foods they have recently consumed. Based upon the assumed composition of the raw materials, the expected losses between slaughter or harvest and eventual human consumption, estimates are made of nutrients available and ingested. These estimates are compared with recommended dietary allowances (RDAs) from standards prescribed by various authorities and regulatory agencies.

The difficulty that inevitably arises is to decide what range of nutrient intakes above and below the levels recommended may be considered nutritionally satisfactory. A nutrient intake deemed adequate may range between what is required to maintain a high concentration in a body tissue and the minimum needed to offset clinical or laboratory indications of specific deficiency. As discussed later, it is customary to incorporate in each recommended daily intake, an amount higher than the assessed minimum to act as a safety factor. As is evident from wide disparities in recommended levels of intake, an international concensus on what constitutes an ideally nutritionally balanced diet for all conditions of women, men and infants will likely remain elusive for some time to come.

FOOD SECURITY AND SELF SUFFICIENCY

An indicator of food insecurity in a country or community is the number of people whose food consumption falls below a predetermined minimum level deemed necessary for body maintenance and active good health. First attention is often focussed on energy requirements and on the estimated caloric intake of the total population. A person is considered undernourished if his or her daily calorie intake is below a desirable minimum.

Various objections have been raised to concepts of undernourishment, it being questioned whether energy requirements proposed by certain international agencies are reliable and applicable across the world. Inadequate caloric intake is not the sole criterion of malnutrition. Economists raise such questions as: Can food security for national populations or particular communities be evaluated on the basis of average per capita caloric intake since this overlooks large personal differences caused by inequalities in income and access? Aggregate statistics overlook the extent of food insufficiencies among the most poorly nourished. Gross aggregates tend to ignore temporary shortages in the accessible food supply that aggravate the deprived condition of the already undernourished and vulnerable. Food security must take full account of differences in income distribution (Sen 1981).

FAO estimates that because of poverty and related disadvantages, over 800 million people suffer nutritional deficiencies; without radical change in economic conditions and political dispositions the extent of food insecurity will worsen. By any moral and civilised standards it is unjust that so many are allowed to suffer and die prematurely while others enjoy a vast abundance. Science can, if so directed, contribute significantly to the alleviation of distress from malnutrition, but the many inequities and insecurities cannot be remedied by science alone.

Causes and Analyses of Insecurity

The most evident food insecurities arise from drought, other climatic disruptions, and political strife. Less publicised chronic malnutrition comes from poverty exacerbated by inadequate, inequitable distribution among different income groups and communities.

Abbott et al. (1993) define "self-sufficiency" as total supply from domestic production, and "self-reliance" as being in part dependent on agricultural trade, the income from what is sold being enough to cover the cost of what is needed. Transitory shortages and chronic undernutrition seem doomed to persist for many years among the poorest countries and communities. Inadequate supply and unstable markets are the consequences of unsustained indigenous production, inability to import, inefficient distribution and variable demand. Capacity to import is influenced by world prices, foreign exchange reserves and the drain on these reserves imposed by debt servicing. Inefficient transportation services impede transfers from farms to markets, and the distribution of imported commodities. Levels of demand vary with fluctuations in disposable incomes, particularly among the poorest people.

Instability and inconsistency in indigenous production is the most prevalent and least easily remedied cause of food insecurity among poor countries. Poor farmers cannot afford the external inputs needed as technological insurance against unpredictably inconstant climates. Abbott et al. (1993) suggest how government trade and pricing policies may alleviate food insecurity. They recommend direct income subsidies to stimulate demand in preference to subsidies to farmers, and discuss possible measures to stabilise prices. Policies heavily reliant on imports gamble that essential supplies will always be available when needed. Price stabilisation may not guarantee security for all income groups.

Abbott et al. (1993) propose an econometric model for the analysis of alternative food security policies, and conclude that (1) purchasing power exerts a greater influence on food security than availability of supply and price stability; (2) the cost of stockpiling rises with the level of security to be realised; (3) excessive government involvement in agricultural markets depresses private profitability; (4) nations that are not self-sufficient must define policies that ensure rational complementarity between reserve stocks and trade. The authors claim their model can take account of complex marketing and distribution systems but is not globally applicable and must be modified to adapt to differing agro-economic conditions, production systems, trade, prices, income distributions and purchasing power.

GLOBAL CARRYING CAPACITY

"Carrying capacity" is generally conceived as the maximum population that can be supported by production from a particular area of land. Carrying capacity is a term familiar to pasture agronomists and wild life managers who use it as an estimate of the animal population that can be sustained by a particular ecosystem. With modification, it can be indicative of how many humans can be adequately fed over a long period by the land on which they live, assuming no irreversible degradation of the natural resource base. Carrying capacity is an important concept for development economists but, in relation to human populations, the criteria of assessment vary widely among authors. The disparities are most evident among estimates of the population that Planet Earth's resources can sustain. How many grazing animals can be carried on a particular area of natural range land or cultivated pasture, can be determined by scientific experiment and careful observation. If the size of the herd exceeds the productive capacity of the range, animals will either starve or migrate to more abundant pastures.

Long-term carrying capacity for human populations is complex and the complexity magnifies as nations and communities gain in prosperity and diversify from simple subsistence farming to mixed rural and urban economies with food demands beyond the minimum calories needed to keep body and soul together. Carrying capacity can cause confusion between food security and total self-sufficiency. Compared with most other Asian communities, the people of Singapore and Hong Kong enjoy high standards of living, standards unsustainable if they depended on food produced entirely within their own territories. Nations that cannot be agriculturally self-sufficient survive by selling goods and services created by investment, ingenuity and industrious effort.

Populations and Resources

In 1798, Thomas Robert Malthus an English priest published "An essay on the principle of population" in which he argued that the world's population would increase beyond the means of subsistence and that measures to restrict population growth were urgently necessary. In 1803 he further proposed that human morality must exert its influence over population expansion. Since Thomas Malthus made his gloomy prediction, how many people the globe can sustain has been a subject for lively debate and dispute. Tables 1 and 2 present, respectively, different estimates of how many people will in future inhabit the earth, and how many the earth can sustain. Observers with access, presumably, to the same sources of information arrive at widely different estimates of the globe's carrying capacity. Westin (1981) and Mann (1981) contend that the earth cannot support more than 2 billion and argue persuasively for drastic reductions in population growth. Kahn (1982) believes available natural resources aided by scientific ingenuity will be sufficient to support persistent population growth for many centuries to come.

Table 1. World Population Growth.[a]

Year	Million	Year	Billion
7000 BC	5-10	1997	6
01 AD	200-400	2025	8
1850	1000	2100	5, 8, 12, 16, ?
1930	2000		
1960	3000		
1975	4000		
1987	5000		

[a] Adapted from Lutz and Prinz (1991); Hulse (1993).

Table 2. Earth's Carrying Capacity.

Author	Date	Number (billion)
Brown	1954	50
Revelle	1976	40
Clark	1958	12
Gilland	1983	7.5
Mann	1982	2.0
Kahn	1982	Unlimited

Some pessimists base their estimates on the assumption that all the world will aspire to and adopt consumption levels equivalent to those of the 27 richest nations. Some optimists assume the earth's natural resource base can be rapidly expanded by technological innovation, and that market forces will keep natural resource supply and demand in satisfactory balance. History shows that as peoples' disposable incomes rise demand for goods and services expand. In making his prediction Malthus expressed the opinion that there should never be more people than could enjoy meat and wine with every dinner.

Globally, Malthus' melancholy predictions have not come to pass. Since 1950 on every continent but Africa food production has risen faster than population, and among industrial countries average grain yields have more than doubled. Since 1960, across the developing regions average grain yields have risen more than 75%. Yet many millions suffer chronic hunger and malnutrition. Thus, the total food calories harvested divided by estimated global population is a reliable indicator of neither the planet's carrying capacity nor of global food security.

The huge discrepancies between the optimists and pessimists are many but, simply stated, are the result of different assumptions of (a) the rate at which science and technology can expand sustainable resources and carrying capacity, (b) the extent to which political, social, economic and market mechanisms will control supply, demand and efficient use of natural resources, (c) what will be the global aggregates of per capita consumption of all essential renewable and non-renewable resources. Since none of these can be predicted with any certainty, it is unlikely that global carrying capacity, the size of world population that can be sustained long into the future can ever be determined. In point of fact the assessment of global carrying capacity is never likely to be more than a diverting academic exercise.

In related discussions, there tends to be confusion between how large the global population might become, and to what size it can safely be allowed to grow. Differences arise between those who assume an ongoing status quo, in which a minority enjoy far greater access to essential resources than the poorer majority; and those whose predictions are predicated on equitable access among all nations and communities. Some ecologies being demonstrably able to grow and harvest more food per hectare of land than others, those who seek equity assume a willingness of the more favorably endowed to share their surpluses with those less privileged, an assumption that neither past nor recent history would seem to justify. Optimistic forecasts also anticipate a technological and logistical capability to distribute fruits of the earth uniformly, and for all people to use resources with comparable efficiency.

In discussions of development in general and of carrying capacity in particular, concerns about population control often fail to distinguish between population size within regions, and rates of population growth. Where resources may be sufficient to sustain an existing population, excessive rates of population growth through high birth rates, greater longevity, and massive immigration, can cause serious economic, logistic and ecological dislocations. Population size and population density are neither the sole criteria of carrying capacity nor the only causes of ecological degradation. Low population densities exist where the productivity of the resource base is low. Asian croplands abundantly watered by monsoon rains support far higher population densities than drought-prone areas of Africa and the Middle East. Two or more crops per year can be harvested from monsoonal Asia, while semi-arid Africa can support at best one meagre grain crop, or only nomadic pastoralism. Large areas of Africa with higher precipitation are rendered uninhabitable by the tsetse fly and other vectors of chronic diseases that afflict humans and livestock.

McCalla (1994), in his Crawford Memorial Lecture, discusses divergent views on global production capacity, classifed as "The Conventional", "The Optimists" and "The Pessimists".

The Conventional perceive a need to double production on presently farmed land without detriment to the resource base. Implicitly, nations will be responsible for their own food security.

The Optimists believe there are many options by which to satisfy future demand, which will be made easier if present production trends continue. The prospects for the period 1990 to 2010 are very good.

The Pessimists refer to falling rates of growth in grain production; fish production from natural sources has reached its biological limit; the livestock carrying capacity of rangelands has been exceeded; growing human demands are pressing close to the limits of water, fish and livestock supply; social disintegrations, population growth and environmental degration are undermining abilities to expand food production.

McCalla concludes that the global community needs a balanced, reasonable analysis upon which to base future decisions.

Since concepts of global carrying capacity are both contentious and contradictory, it would seem more rational to consider the balance and interactions between resources and population by particular regions, communities and political jurisdictions. Even within these there can be confusion between those who equate food security with self-sufficiency and those who, remembering Singapore and Hong Kong, recognise that food security may be realised by generating marketable goods and services that can be exchanged for food. Where land and water are limited it may be economic to cultivate exportable high value crops, such as flowers and spices, that yield higher returns to land, water and labour than food grains and root crops.

Consequently, carrying capacity may be better considered in terms of economically efficient use of land and other resources not simply in terms of an indigenous capacity to raise subsistence food crops. Farmers who cultivate profitable cash crops can afford to adopt novel, more productive technologies, involving risks to which poor subsistence farmers cannot expose themselves.

As was the case during the 1970's and 80's, price trends may indicate temporary abundance or scarcity. Yet over the past 100 years the real price of wheat has declined: in constant dollars the price per tonne of wheat in the late 1980's was barely half the price in the 1880's.

While disputes over global carrying capacity are beyond resolution and, for the most part, futile, it is the responsiblity of every national and communal jurisdiction to assess the resources at its disposal, the present and predicted size and composition of the population to be sustained, then formulate policies and programs to ensure food security for all its citizens. Such policies must comprehend technologies and economics of production together with the resources, infrastructures and logistics essential to preservation and distribution.

COMPONENTS OF FOOD SECURITY

The cultivation of crops and husbandry of livestock began about 10000 BC, before when paleolithic hunter-gatherers survived on the meat of wild animals, and the seeds, fruits and roots of naturally occurring plants. Relative to contemporary American diets, those of successful hunters would likely be more than adequate in protein and high in fibre. The fat content would likely vary among regions, wild animals in cold climates putting down substantial adipose tissue both as a means of insulation and a reserve food source for the winter. Migrant wild animals in tropical climates tend to be less adipose. Until primitive peoples learned how to preserve and store foods and thereby carry over surpluses from seasons of abundance to those of scarcity, they would likely suffer cycles of over-eating and hunger, of glut and famine.

From the time that settled farming with crop cultivation and animal husbandry became established until the industrial revolution 200 years ago, food security was closely synonymous with food self-sufficiency: nations and communities subsisted and survived mainly on what they themselves produced. Progressive urbanisation, the migration of people from rural agriculture to industrial towns, stimulated the development of agro-industries, industries that converted agricultural raw materials into preserved and processed foods, to fibre for clothes, domestic structures and furnishings. Agro-industries were the engines of early economic development for all the nations of Western Europe and North America, a fact often overlooked by development agencies. Agro-industries are essential to food security in providing the means to efficient preservation, storage and distribution; technologies that add value and offer gainful employment for men and women both highly skilled and modestly skilled. Establishment of agro-industries close to sites of production and harvest, restrict spoilage and serve to discourage rural to urban migrations.

Population Versus Production and Distribution

In simplist terms, food security is the outcome of a dynamic balance between demand and supply: the production of food crops, livestock and fisheries being progressively expanded to satisfy rising and shifting needs and demands of growing populations with increasing disposable incomes. Table 1 illustrates the rate at which the world population has grown together with various estimates of its future dimensions. In absence of a cataclysmic catastrophe it seems probable that global population will reach 8 billion by 2025. What will happen during the remainder of the century is uncertain, as illustrated by the many widely disparate predictions. Forecasting how many people the planet can sustain on the basis of production possibilities seems a futile exercise since the outcome depends more on political decisions and economic factors than on science and technology: on how equitably wealth and food resources are distributed and made accessible.

Table 3 indicates how demographic patterns have changed since the late 1960's. In 1967 the more affluent nations were home to 34% of all people and controlled close to 90% of effective wealth. By 1990, the wealthiest minority, 16% of people, controlled 82% of global GNP, with an average income 50 times the income of the poorest 61%. During the intervening period a new group of middle income newly industrialised countries was expanding, largely in east Asia and Latin America, with considerable implications for the global economy and future patterns of food demand.

Table 3. Demographic Distribution.

1967	Industralized		Developing
Population (% World)	34		66
GNP (% World)	88		12
1990	**High Income**	**Low Income**	**Mid Income**
Avg. Income ($USD)	17,500	320	1,900
Population (% World)	16	61	23
GNP (% World)	82	6	12

Table 4. Population and GDP Growth.

	Population (Billion)		Growth (% PA)			GDP Growth (% PA)			
						Total		Per capita	
	1989	2010	1980-90	1990-2000	2000-2010	1980's	1990's	1980's	1990's
World	5.2	7.2	1.8	1.7	1.4				
LDCs	4.0	5.8	2.1	2.0	1.7	3.8	5.3	1.7	3.4
SS Africa	0.5	0.9	3.2	3.3	3.1	2.0	3.7	(1.1)	0.6
N East	0.3	0.5	2.8	2.6	2.2	1.6	4.5	(1.5)	1.6
E Asia	1.6	2.0	1.5	1.5	0.9	8.0	7.3	6.4	5.9
S Asia	1.1	1.7	2.4	2.2	1.8	5.2	5.3	2.9	3.4
L America	0.4	0.6	2.2	1.9	1.6	1.9	3.9	(0.2)	2.1
OECD	1.2	1.4	0.7	0.5	0.4				
E Europe	0.4	0.4	0.8	0.6	0.5	(0.4)	2.1		

SOURCE: FAO (1993); World Bank (1994).

Table 4 presents estimates of regional growth in population and Gross Domestic Product. Despite its low rate of increase, the Asian population which comprises 52% of total, is expected to grow by one billion before the year 2010. It is among these Asian nations that total and per capita GDP and purchasing power are rising fastest.

Demographic Change and Urban Growth

Table 5 shows the rate of GNP increase among several Asian and Latin American countries, several of which, 20 years ago, were looked upon as poor and underdeveloped. The average GNP per capita in India may appear to be low but, it is estimated, at least 250 million in India regularly buy processed foods. Table 6 shows that by 2030 at least one-quarter of the OECD populations will be over 60 years of age while among the nations of Africa, Asia and Latin America the overwhelming majority will be under 30 years of age. Active young people enjoy healthier appetites and eat more than do old folk.

The long-term effects that the accelerating rate of rural to urban migration will impose on food demand supply and security are equally elusive. Table 7 illustrates how fast urban populations are rising in Asia and Latin America. Table 8 reveals that on every continent the proportion employed in productive agriculture is declining and will so continue well into the future. Consequently productivity must increase not only per unit of land, but also per unit of farm labour.

It is impossible to assess accurately the amounts of food produced in rural areas by farm families for their own consumption and sale to nearby communities. During times of stress, rural people may literally live from land to mouth. Food grains, sufficient for a family for a year can be stored in simple silos constructed of materials locally available. Distribution of crops from farms to nearby rural communities is technologically, logistically and economically less complicated than safe delivery of perishable foods from rural to urban markets. Early in the next century over half of the developing nations will live in cities, more than 20 of these cities will accommodate populations in excess of 10 million; 28 cities in Asia will each house over 4 million people.

Food security for inhabitants of sprawling, dense urban communities cannot be provided by subsistence agricultural production. The greater the distance in time and space between the farm and the consumer the more complex become post-production systems of preservation, storage, processing, transportation, distribution and marketing. Far too little attention and too few resources have been invested in post-production systems by development and government agencies. Efficient production and post-production systems are critically complementary and vitally essential to food security.

Table 5.

	GNP/Capita ($USD)		GNP/Capita Growth		Food/Capita Growth (%/yr)
	1982	1991	1960-80	1980-91	1980-91
China	290	370	1.6	7.8	3.0
India	240	330	1.4	3.2	1.6
Malaysia	1620	2520	4.3	2.9	4.1
Thailand	670	1570	4.7	5.9	0.5
Indonesia	430	610	4.0	3.9	2.2
Philippines	690	730	2.8	(1.2)	(1.9)
Costa Rica	1730	1850	3.2	0.7	(0.7)
Chile	2150	2160	1.6	1.6	1.5
Peru	930	1070	1.1	2.4	0.3
Mexico	2090	3030	2.6	(0.5)	0.2
Venezuela	3630	2730	2.6	(1.3)	0
Jamaica	1040	1380	0.6	0	(0.5)

Table 6. Population and age distribution.

	Age distribution: percent over 60 years			
	OECD	Developing	OECD Percent World Population	
1950	11	*ca.* 6	1950	23
1990	17	6.5	2000	12
2030	25		2030	9

OECD: The wealthy nations that are members of the Organisation for Economic and Cooperative Development.

Table 7. Urbanisation.

	Percent of Total Population				Growth (%/yr)
	1960	1970	1980	1991	1980-91
China		18		60	4.5
India	18	20	22	27	3.7
Malaysia	25	27	29	44	4.8
Thailand	13	13	14	23	4.6
Indonesia	15	17	20	31	5.0
Philippines	30	33	37	43	3.7
Chile	68	75	80	86	2.2
Peru	46	57	67	71	3.1
Mexico	51	59	67	73	2.9
Venezuela	67	72	83	85	2.7

Table 8. Proportion of population employed in agriculture.

	1975	1991		1975	1991
World	53	43	M. East/N Africa	54	39
Europe	17	9	Asia	68	59
U.K.	3	2	China	76	67
USA	4	2	India	71	66
Africa	72	63	L America	36	26

SOURCE: FAO State of Food and Agriculture 1992; Production Year Book 1992.

Later in the paper imperfections and shortcomings in infrastructures essential to transportation and distribution from rural farms to urban markets are discussed. There is also reference to studies of rural and urban markets, their multiformity, the critical services they provide, and their comparative neglect by governments and development agencies almost all of whom have devoted more attention and greater investment to agricultural production - to raising yields of crops and livestock, than to the means of protecting, preserving, processing and delivering the products of agriculture in forms that are safe and acceptable to consumers. In an otherwise thoughtful and constructive report on sustainable agriculture the Technical Advisory Committee of the Consultative Group on International Agricultural Research makes only a solitary passing reference to "post-harvest losses" (CGIAR/TAC 1989).

Despite clear evidence that early in the next century over half the developing world's people will live in cities, development agencies and governments seem dominantly devoted to subsistence-scale farming. The importunities of poor subsistence farmers are not to be denied. But they cannot satisfy the nutritional needs of expected millions of future city dwellers. Furthermore, the food and related agro-industries essential to safe food preservation and distribution can provide employment in rural areas and thus discourage migration of the rural unemployed into cities, where facilities and services essential to healthy living are already greatly overburdened.

FOOD SECURITY: ADEQUATE, ACCESSIBLE, AVAILABLE

Berck and Bigman (1993) discuss economic, political and technical issues that bear on food security: on what is available and accessible. Bigman (1993) relates insecurity to chronic undernutrition and temporary food deficiencies, the first reflecting the numbers affected, the second being indicative of the food available and variability in prices. Food security is least uncertain when supply and prices are relatively stable. Bigman comments on indicators of insecurity, diverse concepts of what constitutes an adequate diet and the misleading acceptance of average per capita calorie intake as an index of national nutritional well-being. Average intake estimates conceal individual differences and gaps between actual consumption and minimum nutrient requirements.

Janvry and Sadoulet (1993) contend that sustained economic growth is more effective in reducing poverty than changing income distribution. They disapprove of price mechanism distortions that lead to wastage and misallocation of resources. They emphasise equitable distribution of assets, including land ownership, opportunities for job training and access to health services. Labour-intensive industries provide the means of gainful employment, and thereby relieve pervasive poverty. Agricultural economies in most developing countries are inhibited by unorganised markets and unreliable market information. Neglect of agriculture in general and of a holistic concept that integrates farm production with rural and urban markets constitute serious impediments to economic growth and alleviation of poverty.

Strategic Food Stocks

Santorum and Gray (1993) discuss how public food stocks impinge on food security and contend that stocks of grain are effective against only temporary insecurity.

The primary purpose of food stocks is to provide emergency reserves against global, regional, local and seasonal shortages, to stabilise food prices, overall to counteract fluctuations in supply and availability. Food stocks are usually controlled by marketing boards authorised to maintain sufficient for a defined time period. Minimum reserve stocks are decided either by levels below which the quantities stored must not fall, or residual levels at which action must be taken to replenish stocks. Methods of calculating optimum reserve stocks vary between simple formulae and complex econometric models. Countries with limited ability in planning stocks may opt for criteria defined by other governments, which not infrequently are inappropriate. Minimum reserve stocks, essential in regions subject to frequent but unpredictable episodic drought, in most instances provide only temporary relief and are in no sense a panacea for chronic food insecurity.

Santorum and Gray list various factors to be weighed in determining minimum reserve food stocks. The extent of vulnerability to transitory shortage, caused by drought and other disturbances, varies widely among communities. It is traditional in many cultures to hold on-farm stores sufficient to satisfy the needs of an extended family for at least one year. Vulnerability is dependent on transportation and distribution systems. Communities with easy access to external sources of supply are usually more secure than the landlocked whose internal transportation networks are often unreliable. Massive migrations caused by political disturbances create conditions that no economically sustainable system of food stocks can satisfy. Reported instances of political and military oppressors acting to prevent food from reaching starving refugees present a melancholy reflection of despicable inhumanity.

The effectiveness of reserve stocks in alleviating transitory food shortages depends in large part on an alert early warning system able to signal the probable extent and intensity of an impending shortage. The reliability of early warning systems varies widely among nations. As is discussed later, admirable progress in development of an early warning system is reported by the SADC Food Security Programme in Zimbabwe.

International development agencies would be wise to take account of the vast reservoir of inherited wisdom and experience existent in all rural communities. Several in Africa that as a tradition have maintained reserve stocks, both nationally and on farms, have been persuaded by external agencies to sell grain surpluses to reduce their indebtedness to foreign banks. Sadly, soon after selling their accumulated surpluses these nations suffered severe droughts and crop deficits requiring grain imports that cost considerably more in foreign currency than was realised from the earlier sales.

Food Aid

The World Food Programme (WFP), a United Nations Agency, is the world's leading authority on, and most influential provider of food aid to poor and distressed people (WFP 1993). In recent years roughly $3.5 billion has been invested annually in food aid, close to three-quarters from bilateral, one-quarter from multilateral agencies. Approximately 80% is provided as food aid grants, 20% as food aid loans, total allocations to food aid ranging between 6 and 8% of Official Development Assistance.

Some 14 million tonnes of grain furnish 87% of annual food aid, the balance coming from oils and fats (5%), meat, fish and dairy foods (3%), the residue from various other food commodities.

Food aid, as provided by WFP, serves two essential purposes: (i) relief (ii) development. The first is to relieve immediate urgent distress among people afflicted by war and civil unrest, and others whose normal supplies are destroyed or disturbed by such climatic perturbations as drought, hurricanes, typhoons, inundations and earthquakes. Of some 28 million people who in 1992 received relief food aid, over half were refugees and displaced persons. WFP was responsible for 60% of all international food aid delivered to some 48 countries. Food aid relief was needed in the Horn of Africa because of internecine conflict, and in southern Africa in response to one of the most devastating droughts in the region's history. In consequence of anarchic disruption of ports and normal distribution systems in Somalia and earlier in Ethiopia, WFP was forced to resort to large scale air-lifts to isolated communities. During 1992, for the first time, WFP gave food aid to over 3 million hungry people in the former Yugoslavia.

In addition to mobilising foods from many sources, WFP is required to manage extensive logistic and transportation facilities, by chartering sea-going vessels, fleets of trucks, and in extreme cases aircraft to carry and deliver food by parachute or free-fall.

Relief food aid responds to symptoms and sudden tragedies, but rarely alleviates causes. Though food relief has expanded fastest in recent years, WFPs food aid for development is intended to be of greatest long-term benefit. Food aid for development is given as food for work to promote productive use of human labour; to support basic social services; to provide safety nets for the poor during emergencies and structural adjustment programs. Food aid for development supports labour-intensive agricultural production, food in lieu of wages compensates workers for their time and effort. Food for work has supported a wide diversity of rural activities, including: establishment and maintenance of rural infrastructures; road building and repair; provision of rural schools, health-care and social centres; afforestation and reforestation; land settlement projects to bring new lands under productive cultivation. To improve standards of rural health and nutritional status food aid is embodied in national food reserves and strategic stocks; used to stabilise prices of staple foods, keeping prices to consumers at reasonable levels while offering farmers financial incentives to expand crop and livestock production; to provide food assistance to pregnant and nursing women, to pre-school children; to sustain school feeding programs.

While most past food aid for development has been directed to rural areas, as urban communities expand and urban poverty becomes widespread food for work will be needed to improve urban infrastructures and amenities, to establish and maintain roads and transit systems; to provide clean water, sewage and waste disposal, educational, health, social and other essential services.

Of total food aid for development in 1992, 40% went to sub-Saharan Africa, 10% to North Africa and the Middle East, 32% to Asia and the Pacific, 18% to Latin America and the Caribbean (WFP 1993).

Food aid specifically designed to serve the purposes described is essential and beneficial. Misdirected food aid can cause more harm than good. Such includes indiscriminate dumping of surplus crops and animal products which depress prices and discourage production in Third World nations. Dumping of excessively subsidised livestock products by Europeans has wrought both economic and ecological damage: local markets being depressed, African cattle populations have increased exacerbating soil erosion through overgrazing. Food aid is more sensibly and sensitively administered by WFP than by many self-interested bilateral agencies.

SADC Food Security and Relief

The Southern African Development Community, a cooperative alliance for the members' common good, composed of 11 drought-susceptible southern African nations, maintains a collective food security programme with headquarters in Zimbabwe. Two of SADCs recent developments deserve recognition.

SADCs Ministers have proposed a Revolving Fund from which member nations could borrow to import food when indigenous production is impaired by drought. Borrowers would repay their loans when rains returned and production was restored to normal levels.

A SADC Early Warning System of impending agricultural drought is based upon monitoring of the El Niño-Southern Oscillation phenomenon; the changing patterns of ocean surface barometric pressure in the southern Pacific.

These innovative developments by SADC deserve financial and technical support from donor and development agencies.

CHAPTER 2. FOOD CONSUMPTION

FOOD CONSUMPTION PATTERNS

As discussed below, many factors influence the diets people choose. Food being both a pleasure and a necessity, sensory satisfaction is as much a motivation as physiological need. Foods and preparation methods deemed acceptable derive from ethnic tradition, social culture and acquired habit. Dietary composition and diversity depend on disposable income; eating habits and food choices changing with rising prosperity. Though specific changes are local and personal, certain general patterns of change are widely evident. As incomes rise above what is adequate for bare subsistence, money is spent to bring variety into the diet. Most widely noticeable is for the starchy staple foods of subsistence diets - cereals, roots and tubers - to give way to a higher proportion of livestock products, sweet and fatty foods (Mintz 1989).

Poor people with severely limited disposable incomes cannot afford to be adventurous in the foods they purchase. As their prosperity rises they can buy and try what is interesting but unfamiliar, knowing that if they dislike it they can buy something else. Travel to foreign lands and immigrants from other countries expose people to novel foods and eating habits and stimulate demand for dietary diversity.

Developed Nations

Patterns of dietary change in North America will not necessarily be repeated in other nations. Nonetheless, because of America's wealth and pervasive power, large U.S. food corporations have spread globally to exercise their influence internationally. It is estimated that in 1993 the world ate 95 billion hamburgers: roughly 19 for every man, women and child alive. The global spread of fried chicken and other "fast-foods" is equally apparent. A review of consumption patterns in the USA and Canada may be indicative of changes to be expected elsewhere, since similar kinds of change are already evident in such economically expanding nations of East Asia as Thailand, Taiwan, Hong Kong and Singapore.

United States and Canada

The U.S. NRC Diet and Health Report states that in the USA energy intakes for males from 9 and 64 years of age were between 10 and 17% lower in 1977-78 than in 1965. Over the same period, for females between 23 and 50 years, energy intakes dropped by roughly 9% (NRC 1989). A U.S. National Food Consumption Survey for 1977/78 showed the average daily energy intake for all participants surveyed was approximately 1825 kcal, the group with the highest intake, 2570 kcal, being 15 to 18 year-old males. Men over 75 years averaged 1866 kcal/day. Females with highest average intake, 1850 kcal, were 9 to 11 year olds, lowest being 1420 kcal for women over 75 years. Average intake for women between 19 and 50 years was 1590 kcal.

The proportion of food energy derived from carbohydrates and fats has changed markedly since the start of the century, calories from fats having steadily increased. Energy from carbohydrates declined between 1909 and 1969, but since then has risen by about 9%. Between 1909 and 1985 protein contributed roughly 11%, the percentage of fats rose from 32 to 43%, and the proportion from carbohydrates fell from 57 to 46% of total calories. Changing patterns in average nutrient intake and in food consumption between 1909 and 1985 in the United States are listed in Tables 9 and 10.

Table 9. Average daily per capita nutrient intake USA.

	kcal	Prot (g)	Fat (g)	CHO (g)	Ca (mg)	Fe (mg)	Vit A (IU)	Vit C (mg)
1909	3480	102	125	492	820	15.2	7200	104
1927	3460	95	134	476	860	14.1	7600	106
1948	3230	95	140	403	1000	18.6	8100	114
1965	3150	96	144	372	960	16.7	7300	97
1976	3300	101	157	376	930	18.6	9100	118
1985	3600	102	172	413	920	18.3	9900	114

Though various recent surveys report slightly different average fat intake levels, there is general agreement that over 40% of American citizens' calories are derived from fatty substances, over 42% of total fat being of animal origin. Table 10 compares food consumption between 1909 and 1985, most noticeable being increased intake of beef, poultry, low-fat milk, sugars and syrups, with lower consumption of whole milk, butter and grain products.

Table 10. Food consumption USA (kg/capita/yr).

	1909	1985		1909	1985
Beef	24.5	35.9	Shortening	5.4	10.4
Poultry	8.2	31.8	Salad oils	0.9	11.4
Fish	5.4	8.6	Processed fruits	3.6	15.4
Whole milk	101.2	55.4	Grains	132.1	70.4
Low fat milk	29.1	50.8	Wheat products	98.1	55.4
Cheese	2.3	11.8	Maize products	25.4	3.2
Butter	8.2	2.3	Sugars/Syrups	4.3	69.5

The U.S. Committee on Diet and Health (NRC 1989) summarises results of surveys of food supply, dietary habits, nutrient intakes conducted over many years by U.S. Departments of Agriculture, Health and Human Services, and the National Academy of Sciences. Of particular relevance are the USDA Annual Nationwide Food Consumption Survey Reports (USDA 1984, 1985, 1986, 1987, 1988, 1989, 1990).

The difficulty of determining and documenting actual and habitual intakes of total food and specific nutrients by populations and individuals cannot be overstated. A single 24-hour recall from memory does not offer a reliable estimate of habitual, long-term intake or nutritional status. Estimated nutrient intake is often based on data derived from the disappearance of particular foods into consumers' sources of supply. USDA estimates per capita food consumption from the quotient of total food available divided by total national population, nutrient intake per capita is then calculated from food composition tables. These computations give overall averages but provide little information either on variances among the population, or on quantities and proportions lost by waste, biochemical change and deterioration between farms and consumers' tables. The data simply reflect overall patterns of food availability.

From the data collected, it appears that during this century in the USA, the food energy and protein available per person per day have remained fairly constant at *ca.* 3500 kcal and 100 g, respectively. Most notable percentage increases have been in iron +24, vitamin A +38, vitamin E +57, ascorbic acid +14, thiamin, riboflavin and niacin all *ca.* +37%. Macroscopic surveys indicate that food energy from carbohydrates has fallen from 57 to 46% while calories from fats has moved up from 32 to 43% of mean intake. Fats, sweet foods and beverages high in sugar, contribute together as many calories as vegetables and fruits.

Consumption patterns in Canada have followed trends similar to those in the USA. Early in the century Canadians consumed more than 100 kg/cap of grain products; in 1910 consumption of wheat flour averaged 98 kg. Table 11 shows how wheat flour consumption declined from early in the century until a low of 57 kg in 1985, since when per capita intake has risen to roughly 64 kg with a forecast of increase to more than 70 kg by 2000 AD (Carroll 1958; Agriculture Canada 1990). Favorable acceptance of cereal grain products is stimulated by the advertised nutritional advantages of starch, other "complex carbohydrates", and dietary fibre. Concerns about animal fats and blood cholesterol have inspired the decline in beef, whole milk and egg consumption, while consumption of poultry meat is forecast to double during the last two decades of this century. Similarly, while butter and whole milk have gone down, low-fat milk and salad oils show steady upward inclinations.

Table 11. Food consumption Canada (kg/capita/yr).

	1935-39	1951-55	1980	1990	
Cereals	91	75	65	72	
Meat	63	76	97	96	
Fruit	51	77	61	64	
Vegetables	58	62	54	64	
	1910	1958	1980	1990	2000
Wheat flour (kg/capita/yr)	98	56	59	64	72

Source: Agriculture Canada (1993); Carroll (1958).

Canadian demographic and socio-economic trends and indicators are broadly similar to those in the rest of the western world. Population growth will be slow, the highest percentage lying in the upper age groups (Table 6). The expected outcome is a lower per capita caloric demand, smaller unit food portions, a quest for foods that promise good health and longevity, with avoidance of chronic diseases associated with nutritionally unbalanced or excessive diets. The probability of smaller families, households with fewer members, more women in the labour force, forebode more meals from restaurants, fast food outlets and take-away services, and processed foods that need simple reheating in the home. Food processing and marketing industries will need continually to modify and diversify their range of products. Competition will likely intensify between local and foreign enterprises, particularly in regions where free trade agreements are ratified, and where transportation is not a major constraint.

Table 6 sets out the probable pattern of demographic changes into the next century. It is probable that the proportion of Canadian and other OECD men and women under 35 years will fall, while those over 60 will rise substantially. People over 50 years now represent 25% of the Canadian population, and control over 50% of discretionary spending. In general, these upper age citizens have smaller appetites, demand wider variety and eat more meals in restaurants than younger poorer persons. A public generally better informed about diet and health is raising the demand for fresh fruits and vegetables. Publicity devoted to the possible anti-carcinogenic properties of sulforaphane in broccoli have caused demand for the vegetable to double between 1981 and 1987. Over the same period consumption of fresh fruits rose by over 25% while beef fell by almost 20% (Agriculture Canada 1990).

Over the past 20 years, in constant dollars, Canadian per capita income rose by almost 50%, expenditures on food by barely 20%, average spending on food being only 12% of disposable income. By income group, those in the lowest 20% spend about $29, those in the top 20% close to $40 per week on food. The lowest income group spend 21%, the highest quintile 32% of their food expenditures on meals out of home.

Canadian patterns of consumption and expenditure reflect those in much of the western world: a population increasing faster in age than in numbers, with smaller households, more one-parent families and women in the work force, immigration changing the ethnic mix, with longer healthier lives promised from nutritionally superior diets. Hence, while total caloric demand will stay flat or possibly fall, dominant emphasis will be on what are variously conceived as quality, freshness and convenience. Demand for variety, coupled with expanding international trade, will stimulate greater importation of tropical fruits and vegetables into non-tropical northern countries.

Enlarging capacity to transport fresh fruits and vegetables by air may be beneficial to tropical nations and, given freer trade, to the disadvantage of processed products. Transnational corporations are expanding their activities into the newly industrialised countries of Latin America and East Asia, the more imaginative establishing research and product development centres in these regions, where numbers with disposable income are rapidly growing.

The pursuit of a healthy longevity will provide opportunities for those disposed to seek profit through irresponsibility. The obsession with miracle foods and cures evident in the electronic and print media is prone to exacerbate a latent susceptibility to irrational beliefs in megadoses of whatever appears as the latest *elixir vitae*. In the USA the National Council Against Health Fraud has documented many spurious nutritional claims associated with prevention or cure of heart disease and cancer. Unfortunate folk afflicted with such maladies offer a tempting target and market for outright charlatans and excessively enthusiastic advertisers.

Mediterranean Diets

Keys et al. (1980) reported that among middle-aged men specific death rates from coronary heart disease (CHD) and cancer were lower in Mediterranean countries than in northern Europe, the highest incidence of CHD being in East Finland where 39% of calories came from fat, 24% saturated, 11.9% monounsaturated, and 2.9% polyunsaturated. Lowest CHD was found in Crete and Corfu where 33-36% of calories were from fat: 6.5-7.5% saturated, 18-25% monounsaturated. Buzina (1992) studied diets in Spain, Greece, Portugal and Italy, and found substantial increases in meat, milk and dairy foods together with lower intakes of olive oil. Clinical examinations showed significant upward trends in body mass, blood pressure and serum cholesterol among Mediterranean males.

Developing Regions

Patterns of food consumption and nutritional status are as diverse and dissimilar as may be expected from such heterogeneous populations as exist among regions and nations, among communities within countries as vast as China and India. Furthermore, the cost and resources needed to conduct statistically sound surveys of food and nutrient intakes are greater than most developing nations can afford. Consequently, most relevant data comes from United Nations' and other international and development agencies, much of it derived from national production statistics, sometimes adjusted to take account of post-production loss and uneven distribution. Since assessments of availability, access and consumption depend on approximate gross estimates of production and population size, most data, particularly from the poorest regions, is more realistically considered as indicative than definitive and exact.

While it may be reasonable to believe the planet produces enough to feed many more than are alive today, that the constraints lie only in inequitable distribution, in many territories the logistical means and infrastructures essential to uniform distribution do not exist, nor can the poorest people afford to buy what they need but are unable to produce. Thus, for inestimable millions, the prospect of malnutrition is immediate and chronic. Extrapolations from recent production estimates into predictions and forecasts of future need, supply and consumption should be treated with marked reservation, generous contingency factors being embodied by food and agricultural planning agencies.

From a study of 93 developing countries, FAO estimates that out of a total population of 3.9 billion, some 780 million, 20% of the total, are seriously undernourished, with annual average caloric intakes below what are required to support essential activity and maintain healthy body weight (FAO 1993).

While the global percentage classed as undernourished has fallen since 1970, mainly because of exceptional improvements among several Asian nations, in absolute numbers there has been little change for the better in either south Asia or Latin America. Worst of all, the numbers of malnourished in sub-Saharan Africa have doubled from *ca.* 94 million in 1970 to 180 million in 1990. The proportion of undernourished among sub-Saharan Africans has remained constant at *ca.* 36% during the past 20 years, a chronic condition periodically aggravated by drought and political disturbances. Tables 12, 13, 14, 15 and 16 give statistics of food crop production, degrees of malnutrition and self-sufficiency.

In nations where total production is below levels needed to ensure self-sufficiency, where systems and facilities for storage, transport and distribution restrict nationwide accessibility and adequacy, consumption patterns will vary widely among income groups, between urban and rural dwellers, and among rural people living under different agro-ecological conditions. Energy intakes among Sahelien rural peoples often fall below minimum needs during the months immediately prior to the harvest. Shortages result from depletion of on-farm food grain reserves and the higher energy output in collecting and handling the harvest, preparing the land and planting the seed for the next crop. Energy deficits are revealed in seasonal body weight losses, adverse effects on the nutritional status of pregnant and lactating women, which in turn reflect on the well-being of the children (IDRC 1980; Hulse and Pearson 1980).

Hulse and Pearson (1980) reviewed food consumption and nutrient intake patterns in semi-arid tropical countries of Africa. Recently, USAID (1991) estimated caloric sources intakes by urban and rural populations in Senegal. Mean intake is estimated at 2400 kcal/cap/day close to 75% being from cereal grains, the remainder from vegetables - onions, sweet potatoes, white potatoes and tomatoes being dominant.

Table 12.

	Food Supply (kcal/capita/day)		Undernourished (% of population)		Agricultural Production (Increase %/yr)	
	1970	1990	1970	1990	Total	Per capita
SS Africa	2140	2100	35	37	1.9	(1.1)
NE/N Africa	2380	3010	24	8	3.1	0.3
E Asia	2020	2600	44	16	4.1	2.4
S Asia	2040	2220	34	24	3.1	0.7
L Amer./Carib.	2500	2690	19	13	2.9	0.6

SOURCE: FAO (1993).

Table 13.

	Food Supplies (kcal/capita/day)			Chronic Malnutrition (Millions)		
	1969-71	1988-90	2010	1970	1990	2010
World	2430	2700	2860			
LDCs	2120	2470	2730	940	780	640
SS Africa	2140	2100	2170	95	175	295
M East/N Africa	2380	3010	3120	40	25	30
E Asia	2020	2600	3060	500	250	70
S Asia	2040	2220	2450	250	270	200
L America	2500	2690	2950	55	60	40
Europe	3310	3380	3380			
OECD	3200	3400	3470			

Table 14. World Food Production Indices (1979-81 = 100).

	Food crops		Livestock		Food/Capita	
	1985	1991	1985	1991	1985	1991
World	115	124	112	126	105	103
Africa	115	140	110	133	97	98
M East/N Africa	119	134	120	133	100	96
S Asia	121	141	133	178	111	119
E Asia	122	143	138	194	112	122
China	126	148	152	234	121	137
USA	107	101	107	112	102	93
W Europe	109	109	105	105	106	104
Former USSR	108	97	110	114	105	97

SOURCE: FAO SOFA 1992; Production Year Book 1992.

According to the USAID report on Senegal, consumption of cereal grains varies markedly between urban and rural populations. Among urban homes, rice provides nearly 50%, wheat 25%, pearl millet and sorghum 17%, maize 6%, cowpea around 1%. Among the rural people, millet and sorghum constitute close to 60%, rice 20%, maize 9%, wheat 2%, cowpea 3%. It is not helpful to the economy in general, the balance of trade and foreign exchange reserves in particular, that almost 89% of the rice and all of the wheat consumed are imported. While the West African Rice Development Agency and the Senegal Institute of Agricultural Research have raised rice yields in recent years, rice production has not kept pace with rising demand, particularly in the urban areas (WARDA 1991).

In Senegal, as in other sub-Saharan African countries that straddle several agro-ecological and pluviometric zones, food production and access to food vary widely between northern territories where annual precipitation is uncertain, irregular and averages less than 400 mm, and areas to the south where rainfall may exceed 1000 mm. Barely 8% of land cultivated in Senegal is irrigated and in common with other African nations, inland water resources are not employed as efficiently as they might be (USAID 1991).

Table 15. Cereal Production and Self Sufficiency Ratio.

	Production (Mt)	SSR (%)		Production (Mt)	SSR (%)
	SS Africa			E Asia	
1970	36	97.4	1970	211	97.9
1990	54	86.4	1990	409	96.2
2010	109	85.5	2010	620	96.6
	N East/N Africa			S Asia	
1970	46	86.9	1970	121	97.9
1990	73	65.4	1990	209	96.8
2010	119	62.7	2010	307	96.7
	L America/Caribbean				
1970	66	105.0			
1990	99	87.6			
2010	159	86.5			

Table 16. Food production growth rates: all LDCs (%/yr).

	1961-70	1970-80	1980-90	1990-2010
Cereals	4.1	3.0	2.8	2.0
Roots, plantains, pulses	2.5	1.5	1.6	1.7
Other food crops	3.1	3.5	3.7	2.8
Livestock	3.9	3.8	4.6	3.4

Access to milk, meat and fish is widely variant throughout Africa, the Touareg and other migrant pastoralists subsisting largely on animal protein, while sedentary crop cultivators who live far from coastal and inland waters, subsist on cereals, pulses, plantains, roots and tubers. Though Botswana, Kenya and Zimbabwe maintain efficient livestock production, abattoir management, meat and milk distribution systems, the average meat consumption in Africa south of the Sahara is barely 9 kg per capita per year; the average fish harvest less than 8 kg per capita. Consequently, animal protein is not a major contributer to the diets of most Africans.

The scope of this publication does not permit a comprehensive review of food consumption patterns in all nations and territories. Reliably accurate statistics are not everywhere available. Where such data are acquisitioned and analysed, they appear more often as aggregated data, disaggregated information, particularly as it relates to the poorest communities, being hard to come by.

Table 14 compares estimates of present per capita food supplies with those of 25 years ago across the major developing regions. The table shows the spectacular progress made by east Asian nations in raising agricultural production. In the Near East, south Asia and Latin America, production is keeping slightly ahead of population; in sub-Saharan Africa food production continues to fall behind rates of population growth, failing to keep pace with rising need and demand.

Per capita food supplies in developing countries collectively, have risen from *ca.* 1900 kcal during the early 1960's, to almost 2500 today, over which period developing country populations rose from 2.1 to almost 4 billion. Thirty years ago 1.7 billion people, roughly 80% of LDC populations, inhabited countries with average per capita food supply below 2100 kcal. Recent estimates indicate that fewer than 350 million, about 8.5% of total LDCs, live in countries that average less than 2100 kcal per capita per day. Impressive as this progress has been, close to 800 million, 20% of total developing countries populations, suffer chronic malnutrition. Of the 475 million in sub-Saharan Africa, over 175 million, 37% of all, are reported as chronically undernourished (FAO 1993).

Despite admirable progress made in crop production in Pakistan and India, of 1.2 billion resident in south Asia, some 275 million, 24% of the total, are classed as chronically undernourished. In addition to these deprived millions, as described later, many more suffer specific nutrient deficiencies.

Inequitable Distribution Among and Within Regions

A recent report (FAO 1993) contends that because of inequitable and sometimes inefficient distribution, in countries where apparent food supply is of the order of 2700 kcal per day, at least 10% of population may be expected to be undernourished. The proportion of undernourished can rise to between 15 and 25% when average available calories are in the range of 2200-2500. FAO estimates that some 3.3 billion live in countries with under 2700 kcal, and close to 2 billion in countries with an average below 2500 kcal/cap/day (Table 12).

Apparent caloric sufficiency does not in itself guarantee a healthy nutritional status, particularly where food sources are few and derive mainly from starchy roots and grains. Differing concepts of what make up nutritionally satisfactory diets are described and discussed later. It cannot be overstated that chronic undernutrition and deficiency syndromes are most prevalent where there is severe poverty, and where population numbers are rising faster than agricultural production. Table 4 gives estimates of future growth in population and in Gross Domestic Product for the world and by region. While these data give no measure of variability within regions they indicate in which regions there is discernible economic progress. It is disappointing that during the 1960's several African countries (e.g., Ghana, Nigeria) that now suffer serious economic depression were economically more advanced than several Asian nations (e.g., Korea, Taiwan) whose rates of GDP growth are among the highest in the world.

Comment on Food Consumption Patterns

It cannot be overemphasised that even among the most advanced and prosperous nations, statistics of past, present and predicted per capita food consumption and nutrient intake are little more than indicative best estimates. Nonetheless, the data available clearly illustrate, among affluent and poor nations, that dietary adequacy and diversity are markedly affected by peoples' relative prosperity and poverty; that as disposable incomes rise, patterns of consumption change. Changes evident in North American and Mediterranean diets may indicate what may be expected in the newly industrialised nations: expanding demand for livestock products; declining preference for cereals, pulses and root crops, all foreboding salient implications for agricultural production systems and international trade in agricultural commodities.

Consequently, national jurisdictions responsible for, and agencies devoted to food security planning, prediction and provision must take account of qualitative as well as quantitative indicators in their forecasting methods, whether these be simple arithmetic or complex econometric models.

CHAPTER 3. WHAT CONSTITUTES AN ADEQUATE DIET?

What constitutes an adequate diet for men, women and children around the world has occupied the minds of scientists for many years. The nature and necessity of an adequate diet have been defined in various terms. In Britain during the mid-19th Century dietary standards were laid down as: "the least cost foods to prevent starvation disease during massive unemployment". During WWI food requirements were defined as "what is adequate to feed the army and the nation". Then, as today in many nations, the army had first priority. The League of Nations dietary standards in 1935 were intended: "to marry health and agriculture" (League of Nations 1936). An objective of U.S. dietary recommendations when first formulated was to provide advice "on nutrition problems in connection with national defence".

Published estimates of human nutrient requirements began to be systematically codified during the latter half of the 19th century, though much earlier people had recognised certain foods as having particularly desirable nutritional properties and others that were best avoided (Leitch 1942; Beaton 1991). Foods that were pleasant and acceptable, plant and animal products that were unpleasant, obnoxious, insulting or injurious to health, were discovered empirically and made known to succeeding generations by description and demonstration.

The Technical Commission on Nutrition of the League of Nations, in the mid-1930's, estimated average values for what they perceived as energy, protein and other nutrient requirements of individuals classed by age, sex and activity (League of Nations 1936). Following WWII the newly created United Nations agencies FAO and WHO sponsored a series of expert meetings that recommended dietary intakes each updating and re-estimating nutrient requirements in the light of new knowledge and revised philosophies (FAO/WHO 1962, 1965a, 1965b, 1973, 1988).

Now, at least 50 national, regional and international agencies have formulated dietary recommendations and prescribed desirable intakes of essential nutrients. Notable variations among the many recommendations are reviewed later. Prescribed daily diets are usually set out as specific levels for each nutrient considered essential, each being variously named Recommended Daily Allowance (RDA), Recommended Daily or Dietary Intake (RDI), Recommended Nutrient Intake (RNI). For brevity and convenience, and without prejudice to other designations, RDA will be used in the subsequent text except where specific reference is made to a particular agency's recommendations.

In the beginning, dietary standards were usually estimates of average consumption levels by people considered to be healthy. As scientific understanding of human biology expanded, so did the identification of essential nutrients. The 10th Edition of the U.S. Recommended Dietary Allowances (RDAs) prescribes desirable intake levels for energy and for 24 essential nutrients (FNB/NRC 1989).

Over the past 70 years, estimates of RDAs have progressively been based more on experimental evidence of levels demonstrably necessary to prevent defined deficiency diseases, and less on observed intakes by healthy individuals. Contemporary dietary recommendations take account of subclinical deficiencies and include, as safety factors, amounts in excess of the minimum needed to prevent diagnosable deficiencies.

Beaton (1991) described how an understanding developed of variable individual requirements according to body size and composition, for different levels of occupational and non-occupational activities, the latter recognising that both work and social behaviour impinge on human health and nutritional need. Differing concepts of minimum need and adequate safety factors have led to disparities among various national and international agencies' prescriptions.

From their early advent, one purpose of dietary standards has been to determine quantitatively and qualitatively the total food required for nations and communities. During the 1940's, food planning agencies began to disaggregate gross data to assess the needs of subgroups, households and individuals. Beaton (1991) points to the contradictions in wartime food rationing resulting from such disaggregations: "Of necessity, rationing provided entitlements at the level of each individual but was based (for energy) on per capita population or large group mean requirement estimates".

Beaton distinguishes between two classes of application: diagnostic and prescriptive. "Diagnostic" relates to the examination and interpretation of nutrition survey data and assessment of the adequacy of estimated intake; "prescriptive" relates to planning of food needs and supplies for large populations and groups. The principles and practices of applying present understanding to, respectively, large populations and individuals will be subject to refinement and revision for some time to come. To translate prescribed dietary allowances into information consumers can comprehend and apply when presented on the labels of food packages, calls for concepts of communication and vocabularies less esoteric than those to which nutritionists are generally accustomed. Regulatory agencies in North America and Europe require that food packages carry detailed lists of nutrients, calories and other content for the guidance of consumers.

The medieval quest for an *elixir vitae* has not been forsaken by the sensationalist media, who latch on to every "discovery" that promises immunity to cancer, coronary heart disease and other mortal maladies. Many governments now publish dietary advice in illustrated texts non-specialists can understand. "Canada's Food Guide to Healthy Eating", and Australia's "Dietary Guidelines for Australians" are two such sound, sensible and intelligible publications (H&W Canada 1992; CDH Australia 1983).

RECOMMENDED DIETARY ALLOWANCES

History and Evolution

The evolution of dietary standards has been described by several observers (Leitch 1942; Beaton 1991; IUNS/CABI 1983).

Truswell (IUNS/CABI 1983) comments: "Nutritionists work with foods, about whose composition they have partial knowledge and measure nutrients in food against national standards and recommended dietary intakes (RDIs). Nutritionists in most countries do not have complete sets of composition tables; what they have must be supplemented with data from other countries and with unpublished data. National standards of nutrient intake are often incomplete and sometimes out of date, and differ widely among countries."

The purpose of food standards is to encourage healthy eating habits and to prevent chronic disease attributable to malnutrition. Standards established by the British government in 1862 were intended "to prevent starvation disease" during massive unemployment among workers in the Lancashire cotton industry. It sought to decide: "What is the least cost per mortal for which food can be bought in such quantity and quality as will avert starvation disease in the unemployed population?"

During WWI standards for energy requirements were prescribed in Britain "in order adequately to feed the army and the nation". Standards set down by the British Medical Association in 1933 were designed to maintain health and working capacity during the then economic depression. The earliest recommendations of the U.S. NRC in 1941 were "to maintain perfect health" to realise "a level of (buoyant) health and vigour never before attained or dreamed of".

Thus the early stimulations for dietary standards arose in times of stress and distress: to feed nations in wartime and during economic depressions.

Recommended Dietary Allowances (RDAs) are used for various purposes by governments, international and development agencies: (i) to assess food intake adequacy among selected populations and persons; (ii) to plan national agricultural and food policies; (iii) to balance food nutritional quality against human needs; (iv) to provide adequate food during emergencies for poor and disadvantaged people; (v) to formulate nutrient labeling standards for packaged foods.

RDAs usually embody margins of safety to allow for variable needs among groups and individuals, related to locally available foods and dietary habits. Standards tend to be based upon the level of each essential nutrient that meets the needs of most individuals; RDAs normally exceed minimum physiological requirements to allow for (i) variations among individuals and environments; (ii) conditions of severe stress; (iii) uncertainties of nutritional composition and need; (iv) inefficient nutrient utilisation and inconstant availability.

RDAs were first established by estimating the average physiological requirement for an absorbed nutrient. This estimated value was then modified to compensate for incomplete utilisation and to take account of variations both in requirements among individuals and in the bio-availability of various nutrients among different food sources. This has been described as building a safety factor into the RDAs for each nutrient, a factor which reflects the state of knowledge about the nutrient, its bio-availability, and variations within the population.

Because of uncertainties in methods of determination, differences of opinion exist among nutritionists on the criteria by which to determine physiological requirement for each nutrient. RDAs for infants take account of needs to maintain healthy rates of growth and development; for adults, what will maintain body weight and prevent depletion of all essential nutrients as judged by nutrient balance determinations (FNB/NRC 1989).

Contemporary Criteria

For some nutrients the RDA indicates an intake that will prevent failure of a specific bodily function or manifestation of specific nutrient deficiencies. Ideally, to define an RDA reliably one should begin by defining the average physiological requirement of healthy individuals representative of each relevant age and sex group. Variability among individuals within each group should be assessed so as to define an RDA which satisfies the needs of all healthy members. The difficulties of precise definition are many. Experiments on human beings are expensive and time-consuming, so that only relatively small groups can be studied, added to which, employing humans as experimental animals raises ethical questions in some peoples' minds.

Since few if any population's nutrient requirements follow a normal Gaussian distribution pattern, one cannot simply determine the mean and add two standard deviations in order to cover most people's needs. For example, several studies have illustrated how iron requirements for women are demonstrably skewed (Surgeon-General 1988).

RDAs are based upon foods consumed in normal diets not as pure substances or concentrates. It is therefore necessary to take account of any factors that affect the absorption of particular nutrients and efficiency of utilisation. Some precursors are biochemically converted in vivo into essential nutrients:For example, Vitamin A is derived from certain carotenoids. Digestion and absorption of some nutrients may not be total in which case RDAs must make compensatory allowances. For nutrients which at high levels of intake may be toxic, upper limits of safety must be prescribed. Because of incomplete knowledge, RDAs cannot be prescribed for all nutrients believed essential. Therefore, nutritionists and public health agencies tend to recommend diets based upon a wide variety of available foods (H&W Canada 1992).

Among affluent nations and societies, access to suitable variable diets is rarely problematic: foods that are not locally or seasonally available can be imported from elsewhere. It is among the poorer people of the world that dietary choice is most severely limited and where nutritional deficiencies are widely evident.

Estimates of nutrient requirement come from various sources: from observations on ranges of intake among apparently healthy people; from experimental studies on laboratory animals, and where possible on human subjects. More direct guidance comes from (i) the lowest level of each nutrient intake that alleviates symptoms of a deficiency syndrome; (ii) the lowest intake that maintains metabolic balance over time; (iii) minimum intake to achieve tissue saturation and/or to maintain normal physiological functions.

Because nutrient needs vary among different classes of individuals, RDAs are often given with reference to persons of different age, sex, body size and activity. Special allowances are made for lactating and pregnant women, and for infants on the basis of the average amount of each nutrient ingested by healthy infants when breast-fed by well-nourished mothers. Human milk is generally nutritionally adequate for the first 3-6 months of life, after which supplementation is desirable.

Constraints to Precise Prescription

RDAs are given as desirable levels of intake estimated as average diurnal intakes over time, the period of averaging being dependent on the rate of bodily turn-over of each nutrient. Vitamin A can be stored and used in the body over relatively long periods, whereas others, such as thiamin, turn over more rapidly and deficiency symptoms may appear within weeks rather than months. For most nutrients, if the estimated requirement is not met every day, quantities stored in bodily tissues or consumed subsequently, can compensate for temporary deficits.

Except for people held in a metabolic ward or comparable confinement it is difficult to assess accurately nutrient intake and utilisation. Surveys based on ex post facto recall are at best approximations. Corrections to allow for different ages, environments and activities must also be regarded as best estimates. Since age is not perfectly correlated with health and activity, recommendations according to age must be treated with caution. It was once believed that people working in cold climates needed more food than those in temperate zones. Thirty years ago, packaged rations for men engaged in winter military exercises in northern Canada provided over 5000 kcal per day. It was later realised that with insulated clothing and the need to move slowly to prevent freezing of perspiration, high caloric intake is unnecessary. Highly active people, in whatever environment, need more food energy than those who are sedentary. In tropical arid environments, additional water and salt is needed to compensate for sensitive and insensitive losses.

RDAs are estimated for normal healthy people and do not take account of deficiencies caused by metabolic disorders and chronic diseases. Dietary requirements for particular maladies require specific clinical diagnosis and prescription; they are not addressed in this publication.

Energy Requirements and Food Source Equivalents

A WHO Working Group has defined Energy Requirement as follows: "That level of energy intake from food which will balance energy expenditure when the individual has a body size and composition, and level of physical activity, consistent with long-term good health; and which will allow for the maintenance of economically necessary and socially desirable physical activity. In children and pregnant or lactating women the energy requirement includes the energy needs associated with the deposition of tissues or secretion of milk at rates consistent with good health." (WHO 1985).

Recommended energy allowances normally represent the average needs of individuals, in contrast to RDAs for other nutrients which are usually sufficiently high to account for variability among individuals within populations. If energy intake is consistently higher or lower than a person's requirements changes in body weight and composition will result, either of which may lead to chronic ill-health. Energy allowances and expenditures are normally expressed as kilocalories (kcal) per day. Recommended energy allowances are usually based upon the amount of potential food energy that can be absorbed and utilised. Several authors, including Southgate and Durnin (1970), have determined food energy conversion factors, generally accepted equivalents being: 4 kcal per gram of food protein or food carbohydrate; 9 kcal per gram of food lipid; 7 kcal per gram of ethanol. These are approximates that may vary according to the composition and condition of the food ingested.

Total energy expenditure is the sum of energy expended at rest, in physical activity, and thermogenesis, each in turn affected by age, sex, body size and composition, and physiological state. A WHO Working Group (WHO 1985) has set out the equations from which to derive resting energy expenditures (REE), and the factors by which REE must be multiplied to arrive at approximate energy expenditures for various levels of activity ranging from "Resting" and "Very light" to "Heavy", the latter being such as are incurred during intense manual work and energetic field sports. Levels of recommended energy intake vary with age, sex, body weight and height, and levels of activity.

In assessing nutritional needs, carbohydrates and dietary fibre may be considered together. Sugars include glucose, fructose, sucrose, maltose and lactose; complex carbohydrates include starches and dietary fibre. Dietary fibre includes indigestible complex carbohydrates present in plant cell walls, various gums and mucilages, lignin combined with cellulose also present in plant cell walls. Carbohydrates as sugars and starches are the main source of energy for all mankind, the proportions varying, in some degree, according to affluence and disposable income. Apart from obligatory vegetarians, it is customary that as disposable incomes rise, the proportion of starchy foods eaten declines while energy from protein, particularly animal protein, and lipids increases. Adults in the USA derive roughly 45% of their energy from carbohydrates, about one-quarter of total energy from sugars. Since 1970, in the USA the consumption of fructose produced by enzymatic conversion of glucose from maize starch has risen rapidly (NRC 1989).

Protein

Proteins are continuously broken down and resynthesised, much more protein being recycled in the body than is normally ingested. Economy of protein metabolism depends heavily on the reutilisation of amino acids in the body's biochemical processes, some amino acids are lost by oxidative catabolism, other products of metabolised amino acids are excreted in urine, faeces, other body secretions and in lost skin, hair and nails. Dietary amino acids are needed in continuous supply to replace these recurrent losses. Nine amino acids: histidine, leucine, isoleucine, lysine, methionine, threonine, phenylalanine, tryptophan and valine are not synthesised by mammals and must be included in human diets. These are therefore known as essential amino acids (EAA).

Protein deficiency, at one time believed to be the most serious of deficiencies in Third World countries, is rarely an isolated condition but most often accompanies energy and other nutrient deficiencies that result from an insufficient overall food intake. Symptoms of serious protein deficiency appear as stunting, edema, thin and fragile hair, and such biochemical upsets as low serum albumin and hormonal imbalance. A WHO working group recommended that adult protein allowances be based on nitrogen balance studies (WHO 1985). A United Nations University working group describes techniques and methods for protein quality evaluation and their underlying principles (Pellett and Young 1980).

Protein synthesis and breakdown are sensitive to inadequate energy intake, determination of nitrogen balance being influenced by the body's energy balance. The difference between nitrogen intake and the amount excreted in urine, faeces and other secretions, is known as nitrogen balance. With methodologies available and other inherent constraints, it is difficult to make precise measurements of nitrogen intake, output and balance (Torun et al. 1981; Pellett and Young 1980).

The human body can adapt to different nitrogen intakes. Levels of intake marginally sufficient to sustain growth and to compensate for normal losses, differ markedly from intakes so low as to cause serious deficiency syndromes. In general, protein requirements for children are more broadly agreed upon than for adults. EAA requirements for infants are based on growth and nitrogen accretion; for adults the requirement is taken as the intake needed to achieve nitrogen equilibrium over relatively short periods of time (WHO 1985). While it is accepted that, per kilogram of body weight, requirements for adults are lower than for children, exact EAA requirements for adults remain contentious.

RDAs for protein are customarily based on average requirements for high quality Reference Proteins derived from eggs, meat, milk or fish.

UNITED STATES: RECOMMENDED DIETARY ALLOWANCES (RDAs)

While not proposing they be exclusively adopted by all jurisdictions, the United States' RDAs illustrate the issues to be considered in formulating specific dietary recommendations. Since 1941, ten editions of RDAs have been promulgated by the U.S. Food and Nutrition Board. RDAs are the levels of intake of essential nutrients that, based on scientific knowledge, are judged by the U.S. Nutrition Board to be adequate to meet the known nutrient needs of practically all healthy persons (FNB/NRC 1989).

The 1st Edition of U.S. RDAs was published in 1943 during WWII with the objective of "providing standards to serve as a goal for good nutrition". It recommended daily allowances of nutrients believed essential for people of different ages. The original application was to serve as a guide for nutrition problems related to national defence. Subsequently, RDAs have come to serve other purposes: for the planning and procuring of food supplies for population subgroups; for interpreting food consumption records of populations and individuals; for establishing standards for food assistance programs; for evaluating adequacies of food supplies in meeting national nutritional needs; for designing nutrition education programs; for developing new products in industry; to establish guidelines for nutritional labelling of foods.

In the 10th Edition, RDAs are defined as: "the levels of intake of essential nutrients that, on the basis of scientific knowledge, are judged by the Food and Nutrition Board to be adequate to meet the known nutrients needs of practically all healthy persons."

U.S. RDAs are based on various sources of evidence including:nutrient balance studies; biochemical measurements of tissue saturation; nutrient intakes of apparently healthy adults and infants; epidemiological data; extrapolation from animal experiments (FNB/NRC 1989).

Energy and Protein

Energy allowances are arrived at by multiplying FAOs estimates of resting energy expenditure by an activity factor varying from light to moderate activity. Energy allowances range from 2300 to 2900 kcal for adult men, from 1900 to 2200 for adult women.

Protein allowances are based on nitrogen balance studies and rely on the recommendations of WHO and FAO. From analysis of data derived from long-term and short-term N-balance studies, WHO recommends a daily intake of 0.75 g protein per kg of body weight for young male and elderly adults, with higher levels for women during pregnancy and lactation (WHO 1985). U.S. RDAs are closely similar: 2.2 g/kg for infants up to 6 months; 1.0 g/kg up to 14 years; all other adults 0.8 g/kg with supplements during pregnancy and lactation.

Though there is little evidence to suggest that vigorous activity imposes a significantly higher than normal protein demand, there is persuasive evidence of need for extra protein as infants recuperate from various debilitating infections (WHO 1985).

U.S. RDAs are based upon high quality, highly digestible Reference Proteins, adjustments being made for diets high in protein of low digestibility and/or deficient in one or more essential amino acids.

Vitamins and Minerals

Fat soluble vitamins

Vitamin A is comprised of a family of chemically related compounds. Retinol, retinaldehyde and retinoic acid, along with several synthetic analogues, are collectively designated as retinoids. Quantitatively and qualitatively they vary in biological activity. Nutritional needs may be satisfied by ingestion of active preformed retinoids from animal products, or of carotenoid precursors present in various plants and some animal fats.

Of some 500 known natural carotenoids, only 50 are provitamins in that they are precursors of retinol. Absorption and the patterns of transformation of retinoids and their derivatives are well described in Recommended Dietary Allowances (FNB/NRC 1989).

The convention of expressing vitamin A activity in International Units (IU) does not wholly take account of differences in absorption and bioavailability among retinoids and precursors. Based on rat growth data, one IU of vitamin A activity is equivalent to 0.3 μg of all-trans retinol, or to 0.6 μg of all-trans beta carotene. Because of inferior absorption from intestines, 6 μg of beta-carotene is considered nutritionally equivalent to 1 μg of retinol. Vitamin A activity is now more often expressed in Retinol Equivalents (RE), one RE being biologically equal to 1 μg of all-trans retinol, 6 μg of all-trans beta-carotene, or 12 μg of other provitamin carotenoids. Liver, fish liver oils, whole milk and eggs are sources of preformed retinol; carrots and green leafy vegetables of biologically active carotenoids.

U.S. RDA for adult men is 1000 RE, for women 800 RE, with supplements during lactation and pregnancy. RDAs for children 1-3 years are 400 RE, 4-6 years

500 RE, 7-10 years 700 RE. Symptoms of toxicity from excessive intake are evident at 15,000 μg of retinol in adults, 6000 μg in infants.

Vitamin D is synthesised from precursors in skin exposed to ultra-violet light. Since degree of exposure to sunlight varies widely among people and climates, and frequent exposure is associated with skin cancer, Vitamin D per se is considered an essential dietary nutrient. It occurs as cholecalciferol or ergocalciferol and is usually quoted as μg cholecalciferol. The U.S. RDA is 5-10 μg for adults and 10 μg for children aged over 6 months and for lactating and pregnant women.

Vitamin E occurs in plants in two groups of compounds: tocopherols and tocotrienols, the richest sources being wheat germ and nut oils. U.S. RDAs, in alpha tocopherol equivalents are 10 mg for adult males, adult females 8 mg, infants under 6 months 3 mg, 1-3 years 6 mg, 7-10 years 7 mg.

Vitamin K includes a group of quinone derivatives essential for the formation of prothrombin that helps to regulate blood clotting. U.S. RDAs are 80 and 65 μg, respectively, for adult men and women, for infants under 6 months: 5 μg; 6 months to 1 year: 10 μg; above 1 year of age: 1 μg/kg of body weight.

Water soluble vitamins

Vitamin C (L-ascorbic acid) is an anti-oxidant synthesised by many mammals but not by humans. Dietary deficiency leads ultimately to scurvy, symptoms being weakening of collagenous tissue and extensive capillary haemorrhaging. Vitamin C acts as a co-substrate in the hydroxylation of proline and lysine essential to collagen formation, and conversion of dopamine to norepinephrine. Vitamin C stimulates inorganic iron absorption when the two are ingested together. Absorbed ascorbic acid is present in blood plasma and, as intake increases, the plasma concentration rises to a maximum of 1.2 to 1.5 mg/dL. Up to an intake of roughly 100 mg/day oxalate is the main metabolite excreted in urine. With larger ingestions, ascorbic acid is excreted unchanged.

In the U.S. ascorbic acid is provided mainly by citrus fruits, potatoes and other vegetables, the remainder coming from meat, fish, poultry, eggs, dairy products and enriched foods. Average per capita available rose from 98 mg/day in 1967 to 114 mg/day in 1985 when mean daily intakes by adult men, women and young children were, respectively, 109, 77 and 84 mg/day. RDA is 60 mg for adults, to be supplemented during pregnancy and lactation. Metabolic turn-over is higher among heavy smokers who are advised to ingest at least 100 mg/day.

Thiamin deficiency adversely affects carbohydrate metabolism, which when prolonged leads to beri-beri, manifested in disturbances of the nervous and cardiovascular systems, including mental confusion, anorexia and muscular weakness. Thiamin deficiencies among infants tend to be more severe leading to cardiac failure. Deficiencies are most frequent in diets high in white polished rice or flour from pure cereal endosperm. Best dietary sources include whole cereal grains, brewers yeast, nuts, oilseeds and such organ meats as liver, heart and kidney. U.S. RDAs are 1.5 mg for adult men, 1.1 mg for women, 1.6 mg for pregnant and lactating women. Average intakes are estimated as 1.75, 1.05, and 1.12 mg/day for adult men, women and young children, respectively.

Riboflavin acts in the catalysis of various oxidation-reduction reactions. It is readily absorbed and excreted with its metabolites in urine; riboflavin excreted over a 24 hour period is a reliable index of adequacy. Best sources are meat, poultry, fish, dairy products, whole grain and enriched cereals. RDAs are 1.7 mg for adult men, 1.3 mg for women with supplements during pregnancy and lactation. Average daily intakes in the USA are *ca.* 2.1, 1.34 and 1.57 mg, respectively, for adult men, women and infants.

Human requirements for niacin are partially satisfied by conversion of ingested tryptophan. Niacin is essential to such metabolic processes as fatty acid metabolism and tissue respiration. Extreme persistent deficiency results in pellagra, characterised by dermatitis and inflammation of mucous membranes. Pellagra was widely evident early this century in the southern USA and is common in Asia and Africa. Meats are a good niacin source. In mature cereal grains much of the niacin may be bound in a biologically unavailable form. Mean daily intakes in the USA are estimated as 41 and 27 mg of niacin equivalents for adult men and women, respectively. U.S. NRC RDAs are 19 mg NE for adult men, 15 for adult women with supplements during pregnancy and lactation.

Other water-soluble vitamins:

U.S. RDAs include several nutrients not formally recommended by many other agencies. These include;

Folate: RDAs for adult males 200 μg, for females 180 μg, for infants to age 1 year, 3.6 μg/kg body wt.

Vitamin B12 is supplied by animal products but can be synthesised by various bacteria and fungi. RDA for adults is 2.0 μg with supplements during pregnancy and lactation.

Biotin, present in liver, egg yolk, soy beans and yeast, is synthesised microbiologically in the intestine. In default of definitive data, an intake of 30-100 μg/day is provisionally recommended for adults.

Pantothenic acid is widely distributed in animal tissue, whole grains and legumes. Deficiency symptoms appear as various abnormalities. There is too little data from which to prescribe specific RDAs.

Minerals

The body of an adult contains close to 1.2 kg of calcium, 99% of it in the skeleton being constantly turned over through resorption and reformation. In elderly people resorption moves faster than reformation giving rise to gradual bone deterioration. Absorption is stimulated by vitamin D and in lesser degree by dietary protein. Phytic and oxalic acids sequester calcium, iron and zinc thereby reducing availability of these elements. In the USA over 50% of an average daily intake of *ca.* 740 mg of calcium comes from dairy products. U.S. NRC recommends 1200 mg/day for most adults.

Phosphorus is an essential bone mineral, *ca.* 85% being present in the skeletal mass. The element is involved in various biological processes; the energy for many metabolisms being derived from phosphate bonds in adenosine triphosphate (ATP). Because of its multifarious functions in living organisms, phosphorus is found in most foods, main contributors being cereal grains and foods rich in protein.

Roughly half of U.S. dietary phosphorus comes from milk, meat, poultry and fish, with around 12% from cereals. Average available in U.S. diets is *ca.* 1500 mg per capita per day; mean intakes being *ca.* 1500 and 1000 mg/day for adult men and women, respectively. Given its biological ubiquity phosphorus deficiency is rarely encountered, and exact requirements are difficult to determine. RDAs are 800, 1200 and 800 mg/day for infants, teenagers and adults, respectively; 1200 mg when pregnant or lactating.

Iron is an essential element in haemoglobin, myoglobin and several enzymes. Absorption is influenced by the chemical nature and quantity of the iron ingested. A WHO Expert Committee defined anaemia as iron contents of haemoglobin below 13 g/dL and 12 g/dL, respectively, for adult males and females and below 11 g/dL for pregnant women (WHO 1968).

Principal sources of dietary iron are meat, eggs, vegetables, fortified cereals. Deficiencies in the USA are reported most frequently during four periods of life: (a) between 6 months and 4 years; (b) during early adolescence; (c) in adult females because of menstrual loss; (d) during pregnancy because of demands of the fetus and placenta, and blood loss at the time of delivery. U.S. RDA propose 10 mg/day for children and adult males, 15 mg/day for females, 30 mg/day for pregnant women.

Magnesium is found in many unprocessed foods, of which the richest sources are whole grains, legumes and nuts, over 80% being in embryos and outer seed coats. RDAs prescribed for adults are 4.5 mg/kg body weight which averages 350 mg for men, 200 mg for women. Supplements are recommended during pregnancy and lactation. The RDA for children aged 1-15 years is 6 mg/kg.

RDAs are also prescribed by the U.S. NRC for zinc, iodine and selenium (FNB/NRC 1989).

VARIABILITY IN FOOD COMPOSITION

While RDAs prescribed in terms of analysable nutrients are helpful to professional scientists, to those responsible for a nation's health and welfare, and to food processing industries required to declare contents on package labels, RDAs may have little meaning for the public at large. More helpful are government guides to sensible eating which illustrate the available and acceptable foods that will provide sound nutritional balances for particular needs and lifestyles. However, the ingredient contents of diets recommended in guides to sensible eating are simply indicative since the proportion of each naturally occurring nutrient in all food sources can vary significantly. Composition at time of harvest is influenced by genetic inheritance and conditions of cultivation.

Hulse et al. (1980) cite many references to the influence of genotype, location, fertilizer inputs and irrigation on the nutrient contents of sorghum and various millets. The higest and lowest contents cited are summarised in Table 17. Similar ranges of lowest to highest nutrient contents drawn from many analyses of grain legumes are given in Tables 18 and 19 (Hulse 1991, 1994).

Table 17. Composition of sorghum and pearl millet.

	Sorghum	Pearl Millet
Protein (N × 5.7)	7.1-14.2	7.8-19.3
Lipid (%)	2.4-6.5	1.5-6.8
Carbohydrate	70-90	62-89
Lysine (mg/g N)	71-212	109-297
Calcium (mg/100 g)	11-586	30-62
Phosphorus (mg/100 g)	167-751	248-950
Iron (mg/100 g)	0.9-20.0	1.1-38.0
Thiamin (mg/100 g)	0.24-0.54	0.34-0.41
Niacin (mg/100 g)	2.9-6.4	1.5-3.9
Riboflavin (mg/100 g)	0.1-0.2	0.20-0.24

Data are ranges of means from many results gleaned from review of 1300 relevant published papers. Where necessary, data were recalculated to conform to a common standard, e.g., all are on a dry weight basis. Since methods of analysis were not uniformly consistent and in many instances were not detailed, it is impossible to determine to what extent ranges of variability are real, and in what degree they reflect experimental deviation and/or error. This is also true of Tables 18 and 19.

Table 18. Composition of grain legumes.

	Protein	Crude fibre (%)	Lysine (mg/100 g N)	Methionine
Chickpea	15-29	1.2-13.5	394-437	75-97
Pea	21-33	4.6-7.0	370-600	56-82
Faba bean	23-39	5.0-8.5	391-660	44-57
Lentil	20-31	3.8-4.6	391-471	37-76

Table 19. Composition of chickpea.

	Range		Range
100 seed weight (g)	10.4-36.7	Cooking time (min)	
Protein (%)	14.5-29.0	Whole seed	52-98
Carbohydrate (%)	52.4-71.0	Decorticated	26-40
Lipid (%)	3.8-10.2	Biological value	52-78
		Coeff. of digestibility	76-92

Table 20. Composition of fruits grown in Britain.

	Soluble solids	Total solids	Total sugars	Pectin
Apples	9.5-12.6	12.1-18.4	4.2-9.8	0.5-1.3
Blackberries	7.9-10.4	14.5-23.0	3.3-6.7	0.2-0.9
Blackcurrants	10.0-16.7	17.2-22.4	2.2-8.3	0.6-1.7
Cherries	10.7-14.8	12.3-17.5	6.9-10.6	0.1-0.4
Plums	9.3-17.1	10.4-18.4	3.9-10.3	0.5-1.2
Raspberries	5.4-11.9	10.9-20.7	1.3-7.9	0.4-0.9
Strawberries	5.4-13.6	7.3-16.2	3.2-8.5	0.4-0.8

Ranges of results, percent of whole fruit, from various analyses.
SOURCE: Morris (1951)

Martens and Hlynka (1969) review protein contents of wheat grown in the Prairie provinces of Western Canada over a 40-year period. Mean protein values from 200 growing districts ranged from 9.2 to 17.9%. In 1983, from analyses of over 10,000 wheat samples from across Western Canada, protein ranged from 8.0 to 19.7%, with a mean of 14.1%. In 1985, from over 6000 analysed samples, protein ranged from 8.7 to 19.7% with a mean of 13.4% (GRL 1993).

It is not unusual in food science text books to find protein and other nutrient contents of particular crop species quoted as single seemingly absolute quantities. It is unusual for food science texts that quote nutrient contents to identify the particular crop genotypes under discussion, or to describe where and how they were cultivated.

Conditions That Influence Nutrient Content

Soil chemistry influences the nutrient content of cereals, legumes, fruits and vegetables. Biochemical and physiological condition change progressively before and after harvest. Nutrient content is affected by the degree of maturity at time of harvest, by subsequent conditions of handling and storage, by temperature, relative humidity and atmospheric composition in storage containers. Vitamin C is sensitive to heat and oxygen and is reduced in fruits and vegetables stored under tropical aerobic conditions. Diverse fruit species and genotypes mature at different rates leading to variations in nutrient contents. Vitamin C in papaya increases, in bananas and mangoes it decreases with advancing maturity. In oranges, grapefruit and tangerines vitamin C is higher in immature than in over-ripe fruits. Since the total size and biomass of each fruit increases with maturity, a distinction must be made between nutrient content expressed as percent dry matter and as percent of whole fruit. Fruits picked before full maturity are often firmer and less prone to bruising than over-ripe fruits. Bruising and other physical damage can impair nutrient content.

In vegetables stored at -18°C, vitamin C losses from green beans have been reported to range from 30 to 68%, in lima beans from 39 to 64%, in cauliflower from 40 to 60%. Vitamin C loss during cooking of previously frozen vegetables ranged from 10 to 55% (Harris and Karmas 1975).

Morris (1951) discusses the composition of fruits harvested in Great Britain and the proportional variations of essential constituents among varieties and genotypes. From data collected over several years in fruits grown at different locations Morris tabulates highest and lowest values for components of interest to processors (Table 20).

Though mineral elements are relatively stable many vitamins and other organic nutrients can be reduced in concentration or potency during cooking and other processing technologies. Vitamin A is affected by acids, oxygen, intense light and heat, ascorbic acid and vitamin D by alkalies, heat, oxygen and light, thiamin by alkali, oxygen and heat. Mineral elements can be leached out by cooking water and different authors report losses of iron up to 37%, copper to 27%, zinc to 24%, potassium to 20%, calcium and magnesium each up to 17% (Coultate 1984; Fennema 1985; Harris and Karmas 1975; Morris 1951).

Precision in Food Analysis

Exact and reliable food analysis calls for exceptional skill and experience; analysts must be au courant with advances in the state of the methodological art. The Association of Official Analytical Chemists published the 15th Edition of Official Methods of Analysis in 1990. The two volumes present over 2300 methods of analysis. The AOAC, originally known as the Association of Official Agricultural Chemists, published the first edition in 1920 which states the following: "Analytical methods and procedures included in this volume are those which AOAC members have evaluated and validated through collaborative studies and appropriate similar techniques to give accurate and reproducible analytical results . . . for which the method is stated to apply provided analysis is conducted, as written, by a competent scientist."

It may be questioned whether all of the analytical results quoted in the vast food and agricultural literature, are presented by competent analytical chemists. Following a review of over 1300 published papers on the composition and nutritive value of sorghum and various millets, Hulse et al. (1980) state: "Results quoted for calcium, phosphorus, iron and other inorganic elements must be reviewed with some reservation" a conclusion reached in light of the many different methods employed. Though many analysts used methods recommended by AOAC, at least 14 different methods for calcium, 23 for phosphorus and 17 for iron are reported throughout the text. Furthermore, particularly for calcium and iron, it is often difficult to decide what proportion of that determined by analysis was present as an intrinsic component of the original grain, and how much resulted from contamination before or during cooking (Hulse et al. 1980).

In his Plenary Lecture to Congress VII of the International Union of Food Science and Technology (IUFOST) Bender reflected on a tendency to regard food composition tables as having the accuracy and precision of atomic weight tables and states: "This is far from being so. Apart from the fact that the food under consideration differs from the samples analysed to compile food tables there are enormous problems in the way of obtaining reliable values for representative foods. Differences in methodology, of sampling and among foods themselves give rise to considerable differences in values listed in tables from different authorities." Bender refers to the progress made by INFOODS in bringing "much needed reliability to food composition tables" (Bender 1987).

As recently as 1985, an international consultative group concerned with methods of vitamin assay felt able to recommend methods only for vitamins A, C and E

(Brubacher et al. 1985).

In determining protein nitrogen content, Johann Kjeldahl's method, first published in 1883, remains the standard against which other later methods are compared. The Kjeldahl method determines nitrogen content and assumes that nitrogen represents a constant proportion in each natural protein source. The standard Kjeldahl method based upon the oxidative digestion of all amino acid and amide nitrogen to ammonia, is susceptible to at least 20 inherent sources of error (Williams 1974).

There is a general assumption in many published papers that nitrogen represents 16% by mass of animal and most plant proteins. It has, however, been convincingly demonstrated that the nitrogen content of most cereals and legumes is greater than 17.5% and therefore the nitrogen-to-protein conversion factor for oilseeds, cereals and legumes lies between 5.3 and 5.7 not 6.25 as is so often used (Tkachuk 1977).

It is not always clear whether protein estimates are given on a dry weight or an as-is basis. The discrepancies between different methods of protein calculation can be significant. Assuming on a dry weight basis that nitrogen content determined by Kjeldahl equals 2.50.

$N \times 6.25$ at zero moisture = 15.62% protein
$N \times 5.7$ at zero moisture = 14.25% protein
$N \times 5.7$ at 14% moisture = 12.00% protein

Given the many demonstrable sources of error in protein and other nutrient determinations, it is curious how many RDAs are presented with extreme precision, in some cases to a fraction of a gram of protein. It is equally interesting how many tables of food composition record estimates of nutrients in particular foods as single figures rather than as a probable range of values (McCance and Widdowson 1960).

It would appear highly desirable in departments of food science in colleges and universities, equally in departments of crop and animal sciences, that students be made aware of the complexities that surround food analysis, of the skill and experience required in the preparation and processing of samples, and in the calculation and interpretation of results. The replacement of wet methods by modern physical instruments may have inculcated a belief that analysis is now an automated affair accessible to any interested amateur. Such decidedly is not the case.

DIVERSITY IN RDAs AMONG NATIONS AND AGENCIES

A comparison among recommended daily dietary intakes was compiled by a committee of the International Union of Nutrition Sciences in 1983 (IUNS/CABI 1983). The IUNS review covers 39 countries and regional groupings, and 3 international agencies: FAO, WHO and INCAP. RDAs cover people of different ages, sexes and levels of activity. The nutrients listed differ among nations and agencies, though all propose RDAs for energy, protein, ascorbic acid and iron. The more affluent nations cover a wider spectrum than those classed as "developing". The United States lists energy and 30 essential nutrients; Indonesia and the Philippines prescribe eight, Thailand 9 and the Western Pacific group 11 nutrients plus energy. As illustrated in Tables 21, 22 and 23, RDAs have differed over time within agencies and among agencies, modifications having come about as improved methodologies and new knowledge have appeared. As has been stated, RDAs for the most part are best estimates based upon what are believed to be minimum requirements to prevent diagnosable disease from malnutrition with added safety factors to allow for all manner of contingencies and variabilities. Also, authorities that prescribe RDAs rely on advice of scientific committees, whose membership and opinions change from time to time. Some nations, mostly the affluent, undertake periodic reviews of RDAs; others may remain unchanged over many years.

Variances among agencies arise for many reasons: from differences in scientific opinion, from political and/or economic motivations. It may be advantageous to agricultural or horticultural producers if RDAs for particular nutrients are prescribed at levels higher than necessary, but not so large as to constitute a health hazard. Stimulating a demand for ascorbic acid could benefit growers of citrus crops, encouragement to raise calcium intakes could be profitable to dairy farmers.

Sources of Discrepancies Among Agencies

RDAs are defined following scientific analysis of the information available. Uncertainties for almost all nutrients lead to differences among agencies that can give rise to misinterpretation among people with a limited understanding of their intended purpose and how they are derived. Hegsted (1975) suggests that the two main uses of dietary standards: (a) evaluation of food consumption records, (b) planning of diets and food supplies, cannot be satisfied by a single standard for each nutrient. Hegsted proposes two sets of standards and instructions: (a) evaluation of food intake records should be based on estimates of nutrient needs; (b) planning of diets should take into consideration factors other than estimates of individual nutrient requirements. The objective should be to plan diets acceptable to communities of mixed age, sex and activity since this is the manner in which foods are chosen and meals are put together.

Comparatively few poorer developing countries have formulated their own official RDAs, and some opt for FAO/WHO standards. The effort, expense and experience needed to formulate RDAs, to assess diversity of individual and community consumption, and to compare the data against formulated RDAs exceed the resources of many poorer nations.

Table 21 compares RDAs for different sex and age groups in the USA, UK and Canada and illustrates how the intake levels prescribed have changed since WWII.

While U.S. RDAs for protein and vitamin C have been reduced since 1945, recent UK recommendations for several vitamins and minerals are higher than in 1950 (FNB/NRC 1989; BMA 1950; IFST 1993). Table 22 records how British RNIs for essential vitamins and minerals have been revised over the past 40 years. It is believed that a new revision of British RNIs is being drawn up.

Commentary on RDIs, RDAs, RNIs

It is not surprising that concepts and estimates of what constitute desirable intakes of various nutrients have changed over time and, no doubt, will do so in the future. Estimates of protein requirements for European adults varied from 119 g/day in 1865 to 46.5 g/day in 1889. A 1930's League of Nations Committee recommended not less than one gram of protein per kg of body weight; an FAO/WHO 1971 Expert Committee on Energy and Protein Requirements: 0.57 g per kg; a UN University committee in 1981 and WHO in 1985: 0.6 g/kg; the U.S. NRC in 1989: 0.75 g/kg of body weight (Hulse 1981; League of Nations 1936, 1937; FAO/WHO 1973; UNU 1981; WHO 1985; FNB/NRC 1989). The rationale and reasoning underlying earlier estimates of protein need were reviewed by Scrimshaw (1976).

In light of the changing attitudes and disagreements within and among nations and agencies, uncertainties are inevitable among poorer nations with limited access to nutritional expertise, and inadequate facilities for research and monitoring of food consumption patterns. Table 23 sets out the highest and lowest recommendations for energy and most essential nutrients prescribed by more than 40 nations and agencies. For comparable individuals, the recommended intake of protein varies from 105 to 37 g/day; Vitamin A from 1500 to 600 µg; niacin from 24 to 9 µg; vitamin C from 95 to 30 mg; iron from 24 to 5 mg. For several nutrients, the highest recommendations are from members of the former Soviet bloc: for male adults Poland and the former Federal Republic of Germany recommend highest energy intakes; Czechoslovakia highest for protein; Poland and USSR for Vitamins A and E; USSR for thiamin, also for energy and protein for lactating and pregnant women (IUNS/CABI 1983).

At the time of the IUNS review, few had adopted FAO/WHO recommended intake levels. Nor are the FAO/WHO prescriptions in agreement with those of INCAP, the Nutrition Institute for Central America and Panama.

Disparities among Britons and North Americans

Tables 21 and 22 illustrate changing concepts of nutritional need in the U.S.A and the UK during the post-war years. In both countries, RDAs for protein have fallen sharply since the 1940's; over the same period U.S. RDAs for vitamin C (adults) have dropped from 75 to 60 mg/day, those prescribed in the UK have gone up from 20 to 60 mg.

Table 21. RDAs, RDIs, RNIs Post-WWII: Comparison USA, U.K, and Canada.

	Protein (g/day)					Energy (kcal/day/100)				
	A	B	C	D	E	A	B	C	D	E
Children										
4-6 yr	40	24	56	43	19	16	18	15	17.4	18
7-10 yr	60	28	74	49	26	20	20	20	19.8	19-22
Adolescents										
Male										
13-15	85	45	102	66	49	32	25	27.5	26.4	28
16-20	100	59	130	72	58	38	30	35	28.8	32
Female										
13-15	80	46	102	53	46	26	22	27.5	21.5	22
16-20	75	44	93	53	47	24	22	25	21.5	21
Adults										
Male	70	63	87	72	64	30	29	30	25.1	27
Female	60	50	73	54	54	25	22	25	21.5	21
Pregnant	85	60	102	60	78	25	25	27.5	24	24
Lactating	100	65	111	69	74	30	27	30	27.5	25.5

	Vitamin C (mg/day)					Iron (mg/day)				
	A	B	C	D	E	A	B	C	D	E
Children										
4-6 yr	50	45	15	20	25	8	10	7.6	8	8
7-10 yr	75	60	20	30	40	10	10	10.5	12	8
Adolescents										
Male										
13-15	90	50	30	25	30	15	15	15	15	13
16-20	100	60	30	30	40	15	12	15	10	10
Female										
13-15	80	50	30	30	30	15	15	13.5	12	13
Adults										
Male	75	60	20	30	40	12	10	12	10	9
Female	70	60	20	30	30	12	15	12	12	13
Pregnant	100	70	40	60	40	15	30	15	13	18-23
Lactating	150	95	50	60	55	15	15	15	15	13

RDIs for adults assume moderate levels of physical activity.

Column A = US NRC 1945; Column B = US NRC 1989; Column C = UK 1950; Column D = UK 1979; Column E = Canada 1990.

Table 21a. Comparison of U.S. RDAs with Canadian RNIs.

		Age							
		15-18		19-24		25-50		50+	
		M	F	M	F	M	F	M	F
Protein (g)									
	U.S.	59	44	58	46	63	50	63	50
	Can.	58	47	61	50	64	51	63	54
Vit. A (RE μg/100)									
	U.S.	10	8	10	8	10	8	10	8
	Can.	10	8	10	8	10	8	10	8
Vit. D (μg)									
	U.S.	10	10	10	10	5	5	5	5
	Can.	2.5	2.5	2.5	2.5	2.5	2.5	5	5
Vit. E (mg)									
	U.S.	10	8	10	8	10	8	10	8
	Can.	10	7	10	7	9	6	7	6
Vit. C (mg)									
	U.S.	60	60	60	60	60	60	60	60
	Can.	40	30	40	30	40	30	40	30
Thiamin (mg)									
	U.S.	1.5	1.1	1.5	1.1	1.5	1.1	1.2	1.0
	Can.	1.3	0.8	1.2	0.8	1.1	0.8	0.9	0.8
Riboflavin (mg)									
	U.S.	1.8	1.3	1.7	1.3	1.7	1.3	1.4	1.2
	Can.	1.6	1.1	1.5	1.1	1.4	1.0	1.2	1.0
Niacin (mg)									
	U.S.	20	15	19	15	19	15	15	13
	Can.	23	15	22	15	19	14	16	14
Folate (μg)									
	U.S.	200	180	200	180	200	180	200	180
	Can.	220	190	220	180	230	185	230	195

Table 21a. (*Concluded.*)

		Age							
		15-18		19-24		25-50		50+	
		M	F	M	F	M	F	M	F
Ca (mg/100)									
	U.S.	1.2	1.2	1.2	1.2	0.8	0.8	0.8	0.8
	Can.	0.9	0.7	0.8	0.7	0.8	0.7	0.8	0.8
P (mg/100)									
	U.S.	1.2	1.2	1.2	1.2	0.8	0.8	0.8	0.8
	Can.	1.0	0.8	1.0	0.85	1.0	0.85	1.0	0.85
Mg (mg)									
	U.S.	400	300	350	280	350	280	350	280
	Can.	230	200	240	200	250	200	250	210
Fe (mg)									
	U.S.	12	15	10	15	10	15	10	10
	Can.	10	12	9	13	9	13	9	8
Zn (mg)									
	U.S.	15	12	15	12	15	12	15	12
	Can.	12	9	12	9	12	9	12	9
I (mg)									
	U.S.	150	150	150	150	150	150	150	150
	Can.	160	160	160	160	160	160	160	160

SOURCES: FNB/NRC (1989); H&W Canada (1990).

Noteworthy, in Table 21a are comparative RDAs and RNIs from the USA and Canada, close neighbours whose scientific communities frequently communicate and meet at scientific conferences, and have access to the same scientific literature. Though their agro-ecologies differ, each is so prosperous as to have access to a comparable variety of foods. Their food processing industries operate similar technologies, distribution and marketing systems. Yet for several nutrients, notably vitamins C, D, and mineral elements for young people, the two sets of recommendations are significantly divergent.

Curious indeed is the precision with which the two nations present protein recommendations. Whereas in earlier years the USA proposed protein intakes to the nearest 5 g, Canada and the USA now present such meticulous numbers as 44, 46, 51, 58, 61 and 63 g per day. Does the protein requirement for Canadian men truly drop by one gram per day when they turn 50? The protein RDAs are reminiscent of North American airline time-tables which record that aircraft will depart at 09:03 and arrive at 11:17, when frequent flyers know they will be lucky to be airborne by 09:15 and to reach their destination before 11:45.

RDAs, RDIs and RNIs probably convey little meaning to the general public and one wonders whether even health professionals can take practical account of differences of one gram of protein per day. Few indeed are the analysts who can demonstrate a consistent accuracy of this order of magnitude.

RDAs: Indicative not Definitive

Though the numbers may suggest a high order of scientific precision RDAs are simply indicative not definitive of relative need. A Canadian publication "Nutrition Recommendations" (H&W Canada 1990) states: "The primary use of the 'Nutrition Recommendations' is to form the basis for advice to the public on the choice of healthy diets. They should also be used by those responsible for planning and providing the national food supply, to facilitate public adherence by making appropriate foods available. It would be a misuse to conclude that individuals who do not adhere to each (of the recommendations) are doomed to suffer the associated chronic disease, or that embracing a specific recommendation will ensure avoidance of chronic diseases." The authors go on to say: "Compilations of Recommended Nutrient Intakes (RNIs) have been misused by describing intakes of less than that recommended as deficient or inadequate. RNIs are intended to meet the needs of almost all individuals and so exceed the actual requirements of almost all. . . An intake below that recommended cannot be described as inadequate but . . . the more that intake falls below the RNI the greater the probability of inadequacy. Most emphatically it is not possible to judge the nutritional status of an individual solely on the basis of dietary intake."

Table 22. U.K. RDIs for vitamins and minerals.

	1950	1979	1993
Vitamin A (RE μg)	1500	750	800
Vitamin D (ug)	10	10	5
Vitamin C (mg)	20	30	60
Thiamin (mg)	1.1	0.9-1.3	1.4
Riboflavin (mg)	1.8	1.3-1.6	1.6
Niacin (mg)	12	15-18	18
Calcium (mg)	800	500	800
Iron (mg)	12	10-12	14

Table 23. Comparison of RDAs from 41 countries and agencies.

		Highest	Lowest	Mean
Energy (kcal) (41)[a]	M[b]	3200	2500	2840
	F[b]	3000	2030	2417
Protein (g) (41)	M	105	37	66
	F	96	31	58
Vitamin A (μg RE) (41)	M	1500	600	910
	F	1500	450	860
Vitamin D (μg) (31)[c]	M	10	2.5	4.4
	F	10	2.5	5.5
Vitamin C (mg) (41)	M	95	30	47
	F	85	25	47
Thiamin (mg) (41)	M	1.8	1.0	1.3
	F	1.7	0.8	1.1
Riboflavin (mg) (41)	M	2.4	1.3	1.65
	F	2.3	1.1	1.40
Niacin (mg) (40)	M	24	9.0	18.0
	F	21	13.0	19.5
Iron (mg) (41)	M	24	5	10
	F	35	12	17
Calcium (g) (41)	M	1.1	0.4	0.6
	F	1.5	0.5	0.8

[a]Numbers in parentheses are total respondents.
[b]M = Moderately active adult males; F = Females 13-15 years.
[c] N.B. Most agencies opted for either 10 or 2.5 μg Vitamin D.
SOURCE: IUNS/CABI (1983).

It is evident in recent publications of RDAs, RDIs, RNIs, and in the reports of the U.S. Surgeon-General and the U.S. NRC Committee on Diet and Health that among the affluent nations and societies emphasis upon adequacy has largely given way to promoting diets that militate against chronic diseases and physiological dysfunctions causally related to excessive or insufficient ingestion of one or more components of regular diets.

National Guides to Food and Nutrition

Each national food guide should be distinctive and reflective of culturally influenced preferences and what foods are accessible to most people. There may be scientifically sound reasons why particular RDAs in some countries differ from those in others. For example, people frequently exposed to sunlight may need a lower dietary intake of Vitamin D than folk upon whom the sun rarely shines. In the industrial towns of Britain, from the time of the industrial revolution until free milk was introduced into primary schools during the 1930's, rickets were endemic among poor children, the sun's rays finding it difficult to cut a path through the smog generated from the soft coal burned in homes and the dark satanic industrial mills.

Overall, however, one might hope that nutritional scientists the world over will collaborate with international agencies to formulate RDAs that may be sensibly, safely and economically adopted by the poorer developing nations, where malnutrition and food insecurity are most seriously apparent.

People who are committed to eating sensibly and to choosing for themselves and their families adequate but not excessive diets, will likely gain more of practically useful advice from illustrated food guides than from complex, often contradictory RDAs. Canada's "Food Guide to Healthy Eating" is a typically useful example with recommendations that are consistent with the scientifically detailed prescriptions published in the reports of the U.S. Surgeon-General and the U.S. NRC Committee on Diet and Health (H&W Canada 1992; NRC 1989; Surgeon-General 1988).

The Canadian Food Guide identifies foods in four main categories: (a) Grain products, (b) Vegetables and fruits, (c) Milk products, (d) Meat and meat alternatives. It recommends the desirable number of daily servings in each category, a "serving" being defined by weight, volume or other recognisable unit, illustrated by a coloured picture. The Guide advises: (1) choose from a wide variety of foods; (2) give emphasis to cereals, particularly whole-grain products, vegetables and fruits, with preference for leafy dark green vegetables, fruits orange or deep yellow in colour; (3) choose low-fat dairy products, lean meats and other foods prepared with little or no added fat; (4) limit intakes of alcohol, salt and caffeine; (5) maintain a healthy body weight by regular physical activity and avoidance of excessive eating. The Guide lists the essential nutrients provided by each of the food categories. Legumes to provide vegetable protein are advised as periodic alternatives to meat.

Final Comment

RDAs serve many useful functions and deserve serious consideration by all people, particularly those who bear responsibility for national health, food and agricultural policies, and for public health education. Nonetheless, recognising the variability in food composition, the many sources of imprecision in determining nutrient content and intake, one might question whether RDAs need be expressed as precisely as is implied in some recommendations cited above. Poorer developing nations would be helped by provision of simple illustrated guides to healthy eating, based upon the foods that are seasonably available. Particular attention needs to be given to the rapidly expanding urban populations and the logistical and economic difficulties of delivering to them adequately safe and wholesome diets.

PARTICULAR GROUPS AND COMMUNITIES

The Military

Because they live, move and have their being as self-contained discrete entities, food intakes and energy expenditures by soldiers, sailors, air-crews and other military units are more easily determined than nutrient assessments among the public at large. An adequate diet being essential to physical fitness and mental alertness, every department of defence and military command must study need in relation to activity, both in terms of nutritional quality and utility to suit and satisfy a vast diversity of operational, environmental and stressful conditions. Consequently an extensive literature resides in military archives and libraries, much in internal reports, many at one time classified and restricted, a great deal now available to seriously interested persons.

A recent search (DND Canada 1994) brought to light information on the physical and mental capacities of armed service people to tolerate a wide diversity of operational stresses including: the physical performance and caloric balance of commandos with varying carbohydrate intakes during extreme operational activity; nutritional adequacy of different combat rations during sustained field combat operations; body fluid balance among airmen during long periods in the air when engaged in in-flight refuelling; dehydration, energy output and cardiac function at low calorie intake during vigorous field exercises; the effect of stress on fat metabolism; changes in body biochemistry at low caloric intakes; logistical and food requirements under various environmental conditions; delivery of emergency food by air-drop; to mention only a random few. Since the experimental subjects are captive human beings obedient to military discipline, nutritional studies by the military are generally under more precise control than studies on private citizens going about their every day lives.

Starvation and Acute Caloric Deficiency

In a study of survival feeding carried out for the Canadian Defence Research Medical Laboratories, Stevenson (1958) reviews physiological conditions of various people exposed to near starvation and severe caloric deficiency for extended periods. Stevenson states that, under favourable circumstances, a man can survive without water for a week, with water but without food for up to 2 months. To insure against unforeseen isolation from their food source, soldiers, sailors and aircrews carry calorically dense, compact emergency rations. Table 24 cites 1958 recommended intakes for various levels of activity.

Table 24. Recommended Daily Intake.

	Sedentary	Moderate	Heavy	Very Heavy
kcal	35.0-38.0	43.0-49.0	54.0-58.0	65.0-73.0
Protein (g)	1.0-1.1	1.0-1.3	1.0-1.5	1.0-1.8
Fat (g)	1.0-1.1	1.2-1.3	1.6-2.1	2.0-2.6
Carbohydrate (g)	1.0-1.1	1.2-1.3	1.6-2.1	2.0-2.6

All values: per kilogram body weight for healthy adult male. From Stevenson (1958).

During and after World War II extensive effort and resources were invested in research to design and develop effective survival rations for such emergencies as ship wrecks, forced and parachute landings in isolated terrain. Laboratory and field studies of people subjected to severe caloric deficiency provided data from which to formulate food packs suitable for different emergencies.

During starvation the human organism must derive its energy from its own tissues. If the body's carbohydrate store of glycogen is entirely and immediately available, it could supply the body's energy need for about 24 hours. Consequently, to survive for days or weeks, a starving person must draw on the body's store of fat and protein. The records and diaries of starving survivors indicate that hunger is experienced during the first two or three days but, with sufficient drinking water, the hunger sensation abates. As time goes on the victim becomes weaker and apathetic to mental and physical activity. How soon such depressive changes occur depends on the initial state of health, the body reserves of fat and protein, and degree of environmental stress. In a warm or temperate climate, heart rate and body temperature remain near normal until shortly before death (Stevenson 1958).

After the first day or two of starvation, 80-90% of energy output is derived from body fat, 10-20% from muscle protein, vital organs such as the heart, gonads and nervous system being least sacrificed. The basal, resting metabolic rate falls as starvation progresses which allows the organism to economise on its energy stores. Autolysis of muscle tissue leads to lassitude discouraging physical activity.

The elevated rate of body fat metabolism in absence of carbohydrate results in incomplete lipid oxidation causing abnormal production and accumulation of acidic ketones which, to be excreted, require water from the kidneys. Prevention of ketosis during starvation is critical and is assisted by providing digestible carbohydrate in survival food packs. Several authors have published results of laboratory studies on humans subjected to caloric deficiencies of varying severity (Robins 1890; Keys et al. 1950; Benedict et al. 1919).

An experiment, often cited, reported by Benedict (1917) was carried out at the Carnegie Institute. A human volunteer abstained totally from food and drank only water for 31 days. The subject's basal metabolic rate fell steadily throughout the fast from 1441 kcal/h on the first day to 1025 kcal/h on the 30th day. Total nitrogen loss of 277 g was calculated as equivalent to almost 8 kg of protein tissue. In the excreted urine, total solids fell from 43.5 to 27.1 g; sodium from 2.1 to 0.05 g, while beta-oxybutyric acid rose from zero to 4.5 g, indicative of incomplete oxidation of body fat. No impairment of mental faculties was noticed and when examined one year after the experiment, the subject's physical capacity was reported to be normal.

Various reports of chronic starvation have shown body weight losses above 50% of original leading Stevenson to conclude that any "concept of a generally valid lethal level of weight loss must be relegated to the rich store of scientific mythology" (Stevenson 1958).

Stevenson lists several physiological disorders that are typical of starvation: diarrhea, flatulence, non-specific dysentery, bloated abdomen, lower pulse rate and blood pressure, a rapid drop in total metabolism, a more gradual decline in basal metabolism. Stevenson reported that during chronic starvation the magnitude of protein loss varies widely among organs; that plasma proteins decrease with eventual onset of edema. Ketosis, one of the earliest indicators of starvation, is noticed by the sweet odour of acetone on the breath. Incomplete oxidation of fat, that causes ketosis, may be alleviated by feeding small amounts (*ca.* 100-200 g/day) of carbohydrate. Gelatinised starch lozenges are a common component of survival and emergency food packs.

Stevenson found no evidence of specific vitamin deficiency during acute starvation. Urinary excretion of thiamin soon falls to zero; riboflavin excretion rises rapidly during the first few days but then falls. It was several times suggested that a high intake of ascorbic acid was beneficial to survivors, particularly in cold climates, but the experimental evidence is not convincing.

Water deprivation is more quickly debilitating and fatal than lack of food. Stevenson's estimates of body water turnover for a normal healthy man in a temperate climate are shown in Table 25. Extreme dehydration, in arid tropics, sometimes referred to as "desert thirst", passes through three stages: (1) Dryness; the sufferer becomes desperate for water; (2) Functional derangement characterised by "cotton mouth" and, later, shrivelled tongue; (3) Structural degeneration, described as living death with apparent sweating of blood as the skin cracks.

Stevenson describes shipwreck survivors who drank sea water, leading to serious nervous disturbance and delirium. In men deprived of water in the desert, weight loss was rapid, urine volume fell, blood urea rose. In temperate climates a minimum intake of 900 mL/day of free fluid was needed.

Table 25. Body water balance in a healthy man in a temperate climate.

Intake (mL)		Output (mL)	
Free fluid	1200	Vapour from respiration	500
In food	1100	Insensible from skin	500
Oxidation of food	300	Urine	1500
	2600	Faeces	100
			2600

Water Deprivation

Gamble (1944) made an extensive study of water demand during fasting and of experience with castaways. He calculated minimal urine water as 600 mL, insensible loss 1000 mL, giving a total obligatory expenditure of 1600 mL/day during fasting. Water deprivation is especially serious in hot dry environments such as the arid and semi-arid tropics. In trying to maintain body temperature by evaporative cooling, rapid water and salt loss results from heavy perspiration, leading to concentration of extra-cellular body fluids. Replenishment by fresh, salt-free water, has been shown to cause cellular swelling and severe muscle cramps.

Adolph (1947) describes evident symptoms of dehydration exhaustion syndrome: discomfort and oppression, restlessness, lassitude, muscular fatigue, rising body temperature and pulse rate, tingling in the limbs. As water loss increases, dizziness gives way to delirium, absence of salivation to swollen tongue and inability to swallow, lassitude and an inability to walk or speak clearly. Onset of dehydration exhaustion begins at between 5 and 10% deficit of body weight and can be fatal at 25% water deficit in temperate and 15% in tropical climates.

Discussions of food security tend often to ignore the need for water except insofar as it relates to irrigated agriculture and aquaculture. Yet deprivation of water, particularly of clean drinking water is, for many of the world's poorest people, the most serious cause of chronic ill-health. The world's poorest nations exist in arid, semi-arid and drought-prone ecologies where infrequent, uncertain rainfall restricts access to safe drinking water, a condition widely exacerbated by unsanitary disposal of human and domestic waste.

Military and Survival Rations

Military rations are of interest to UN agencies because of the melancholy expanding need for peace-keeping operations.

Every army in history has carried some form of combat or emergency food supply. The ancient Greeks carried a food pack containing squills (roots of a plant of the Lily family), sesame seed, honey and opium, the latter presumably to deaden hunger pains. On other occasions the Greek armies subsisted on dried mallow, oat and legume flour, raisins and dried meat. Various kinds of sun-dried meat - pemmican (North America), biltong (Africa) and charqui (South America) - have provided primitive people and modern military men and explorers with portable, calorically dense emergency foods. Stevenson (1958) gives a comprehensive account of military combat and survival rations from the time of World War I to the time of the Korean war.

At the present time, military men and women not engaged in combat or military exercises are advised to adhere to their national nutritional recommendations, such as the RDAs proposed by ministries of health and national research councils.

In 1987 the North Atlantic Treaty Organisation (NATO) promulgated a Standardization Agreement (STANAG No 2937) which defines and describes minimum nutritional requirements for all types of military rations. Commanding Officers may choose from one of three alternative rations if additional food is deemed desirable:

1) a Light Meal Combat Pack of 1000 kcal
2) an Arctic Supplement of *ca.* 800 kcal
3) a Tropical Supplement of flavoured crystals sufficient to make 680 mL of beverage.

STANAG 2937 defines and describes three specific rations:

1) Survival Ration: For situations where the concern is to stay alive with limited energy expenditure and water supply.
2) Emergency Ration: To sustain operational capability for brief periods when the regular food supply is disrupted.
3) Individual Combat Ration. Each to provide complete and wholesome subsistence for one day. Not expected to be used for more than 30 consecutive days.

The Survival Ration is small in weight and volume and contains at least 100 g carbohydrate as glucose (or polysaccharides that convert to glucose) per person per day. Water intake should be at least 500 mL/day in temperate and 2000 mL/day in hot environments. Water purification devices or tablets should be separately packed. Protein and fat are not recommended. Shelf life required is 4 years at ambient temperatures.

The Emergency Ration, small in volume, should contain concentrated foods to provide at least 1000 kcal/day, 40-70% from carbohydrate, 20-40% from fat, 10-20%

from protein. Foods should be edible and digestible without addition of water or heating. Water purification devices should be separately packaged. A shelf life of 2 years is needed.

The Individual Combat Ration is intended to provide 3200 kcal/day, with 10% from protein, at least half of which should be high biological value animal protein. Fat should provide 35-40% of calories; vitamin and mineral content should be consistent with the recommendations of the U.S. Food and Nutrition Board.

People in Space Vehicles

Men and women living and working in a space capsule provide a rare instance in which physiological and biochemical change in relation to diet and activity can be systematically monitored. Nutritional needs and physiological responses of men and women travelling in space vehicles have been reported in several recent publications (Lane and Rice 1991; Lane et al. 1992, 1993; Lane and Schulz 1992). As exploration expands literally in time and space, new interests and challenges will be presented to physiologists and nutritionists since space flight provides a model for the study of healthy individuals undergoing unique physiological stresses each in a confined controlled environment.

The first man to be fed in space was Yuri Gagarin during a flight lasting 108 minutes in 1961. The Gemini Program (1965-66) extended missions to 2 weeks and imposed heavy physical demands. The Apollo Program (1968-73) flights being of longer duration permitted studies of cardiovascular deconditioning, bone demineralisation and vestibular disturbances. During later Apollo flights energy balance, faecal losses, body weight, body volume and biochemical changes were recorded. The more extended Skylab missions, the longest being over 84 days, expanded the scope to study energy and nutrient requirements and to collect data relevant to cardiovascular, musculoskeletal, haematological, endocrine and other changes in body chemistry.

The Skylab astronauts chose from 70 different frozen, heat-processed and freeze-dried foods in menus planned over 6-day cycles. For those in the 84-day mission, a high calorie food bar was included. During the 84-day mission body weight loss ranged from 0.9 to 3.64 kg, more than half the loss being from the fat-free mass that includes muscle, bone, organs and blood. Total body fluids decreased between 500 and 900 mL.

Long-term exposure to weightlessness caused bone demineralisation particularly in bones that bear most body weight. Bone-mass decreased, urinary and faecal excretion of calcium increased, with a consequent negative calcium balance. Calcium supplements and hormonal treatments did not prevent bone demineralisation. Present concerns are that higher urinary calcium could lead to formation of renal stones and that bone mineral loss could be irreversible.

The nutritional requirements specified by the National Aeronautic and Space Administration (NASA) are:

1) Energy intake shall be sufficient to maintain body weight, body composition at levels of activity demanded. WHO formulae provide the basis of calculation for energy requirements.
2) Between 15 and 20% of calories shall be derived from protein, roughly 50% from carbohydrate, less than 10% from simple sugars, 30-35% from fat, the ratio of polyunsaturated to monounsaturated being 1.0 to 1.5-2.0. Each daily diet will include 10-15 g of dietary fibre.
3) Fluids ingested should be enough to prevent dehydration, reduce incidence of kidney stones, the required intake being estimated at not less than 2000 mL per day.
4) Vitamins A, E and K, B6, B12, riboflavin, folate, niacin and magnesium are as prescribed in U.S. NRC RDAs (FNB/NRC 1989). Intakes of vitamin D higher than proposed in U.S. NRC RDAs are advised to assist retention of calcium since astronauts are deprived of exposure to natural UV light. Vitamin C is raised to 100 mg/day, thiamin to 1.5 mg for men, 1.1 mg for women. Phosphorus is prescribed at 800-1200 mg/d but diurnal intake must not exceed 1.5 times the calcium ingested since a Ca:P ratio above 1.5 is believed to decrease calcium absorption. It is hoped nutritional means of preventing bone mineral loss will be discovered before future missions, far more extended in space and time, are undertaken.

DIETARY BALANCE AND IMBALANCE

The poet Thomas Stearns Eliot asks the rhetorical question: "Where is the wisdom we have lost in knowledge; where is the knowledge we have lost in information?". Eliot might well have had food and nutrition in mind since few other subjects have inspired more people to put pen to paper or fingers to word-processors than those with advice to offer on how to eat and live forever. Among this plethora of information there is a good deal of doubtful and misleading information about painless routes to longevity and food for freedom from all ills. Even some professed scientific literature may be misleading since statistically and biologically reliable research on human nutrition is difficult and expensive to undertake. A particular diet or substance that appears beneficial to one experimental sample may not be so for others of a different disposition or in a different environment. One man's meat may literally prove to be another's venemous poison. ("Quod ali cibus est aliis fuat acre venenum" Lucretius: *De Rerum Natura* iv, 637).

Diets acceptable to any community probably evolved by empiricism, each choosing by trial and error what was found to be acceptable, agreeable and satisfying among the plants and animals locally available. As tribes and individuals moved and migrated, and over the past 50 years as facilities for inter-continental travel expanded, the range of acceptable foods has diversified particularly among affluent people.

Throughout history attitudes to food have been as much influenced by pride, prejudice and superstition as by reason. During the Middle Ages there were those who believed that red wine and red meat promoted good blood and, conversely, that white wine and white meat were conducive to anaemia (Hulse 1981).

An extensive U.S. NRC review of diet and health addresses scientific issues fundamental to nutrition policy in relation to the risks of dietary related diseases. It points to a lack of consensus on the role of diet in the etiology of chronic diseases resulting, in part, from incomplete knowledge and, in part, from any generally accepted criteria on which to interpret available evidence. The report seeks to cross the boundaries that separate simple assessments of single chronic diseases such as atherosclerotic cardiovascular disease, cancer, diabetes, obesity and osteoporosis. The report evaluates data from studies on humans and animals and points out that the effect of diet on chronic diseases has been most consistently demonstrated in comparisons of populations with substantially different dietary practices. More specific details of the report are discussed later (NRC 1989).

Much remains to be learned about effects of diet on chronic disease risk. The U.S. NRC report discusses assessment of risks and benefits and, in particular, what may be the effect of a prescription to relieve one symptom upon other nutritional factors and conditions. For example, recommendations for increased calcium intake to protect against osteoporosis might encourage a sizeable intake of dairy products which could raise the risk of coronary heart disease (CHD).

The general conclusions were that diet can influence the risk of several major chronic diseases, the evidence being highly persuasive for diet related to atherosclerotic cardiovascular diseases and hypertension, and highly suggestive in relation to certain forms of cancer. Several chronic diseases in which nutritional factors may play a role have genetic and environmental determinants, not all of which have been clearly characterised.

REPORTS OF THE U.S. SURGEON-GENERAL AND THE U.S. NATIONAL RESEARCH COUNCIL

Two massive documents published in the United States address issues relevant to the risk of chronic ill-health from nutritional imbalance and excess: (1) "The U.S. Surgeon-General's Report on Diet and Health"; (2) The report of the Committee on Diet and Health of the U.S. National Research Council (NRC 1989).

U.S. Surgeon-General's Report

In the U.S. Surgeon-General's Report on Nutrition and Health (Surgeon- General 1988) the following is in the opening statement: "Message from the Surgeon-General": "This Report on Nutrition and Health follows the tradition of the original report on smoking and health. It addresses an area of controversy and substantial misunderstanding. The relative magnitude of the associated health concerns is comparable, with dietary factors playing a prominent role in five of the 10 leading causes of death for Americans. . . . The depth of the science base underlining its findings is more impressive than for tobacco and health. . . . with animal and clinical evidence adding to the epidemiological studies."

According to the U.S. Surgeon-General many U.S. citizens are changing their diets by consuming products comparatively low in saturated fats. Nonetheless, factors related to coronary heart disease (CHD), high blood pressure, strokes, some forms of cancer, diabetes and obesity all contributing to ill-health, are indicative of the opportunities for improvement in the state of U.S. public health if the U.S. Surgeon-General's recommendations on dietary habits were widely accepted and adopted.

The U.S. Surgeon-General's report is primarily focussed on the relation of diet to chronic diseases and does not emphatically address the more illusory issues relevant to spasmodic or chronic hunger and undernutrition. The report recognises that many Americans either choose not to adopt or lack unstinted access to a suitable diet, both being matters of ongoing concern.

Disease from Excess

The report expresses the need for more reliable monitoring of the nature and extent of hunger, malnutrition and inadequate access, from which to determine underlying causes and remedial opportunities. It makes the point that good health, however defined, depends on complex interactions among various environmental, behavioural, social, genetic and economic factors, some within, some beyond personal control. Though malnutrition and hunger afflict many underprivileged Americans, among the majority eating excessively of diets unbalanced in nutrients is most prevalent. An important conclusion by the U.S. Surgeon-General is that overconsumption of foods high in fat instead of complex carbohydrates and fibre, is of particular concern to U.S. public health officials.

Until the early 1940's, maladies such as rickets, pellagra, goitre and xerophthalmia, resulting from nutritional insufficiencies, were found in various U.S. communities. Today, largely because of fortification with critical trace nutrients, more efficient analytical methods available to regulatory agencies and food industry quality control laboratories, the incidence of micro-deficiency diseases among most Americans has been greatly reduced. Deficiency syndromes have been superseded by disease from dietary excess and nutrient imbalance. The 10 leading causes of mortality in the USA are CHD, cancer, strokes, unintentional injuries, obstructive lung disorders, pneumonia, diabetes mellitus, suicide, chronic liver dysfunctions and atherosclerosis. Five of these: CHD, some cancers, strokes, diabetes and atherosclerosis are demonstrably associated with diet, while liver cirrhosis, accidents and suicide often follow upon excessive alcohol intake.

Costs and Consequences

Over 1.75 million U.S. citizens succumb to heart attacks or strokes every year. The cost in direct health care and lost productivity is of the order of $60 billion. Close to 60 million suffer from hypertension, 34 million are obese, more than 900,000 new cases of cancer are diagnosed every year. The U.S. Surgeon-General's recommendations relevant to these afflictions call for lower intake of cholesterol and saturated fats, with more fruits, vegetables, whole grains, fish, poultry, lean meat and low-fat dairy products in place of fatty foods. Caloric intake should be commensurate with energy expenditure, a balance best realised by choosing low-calorie foods, high in complex carbohydrates and fibre together with a regimen of regular exercise.

Adolescent girls and adult women are advised to take in foods high in calcium; for children, adolescents and women of child-bearing age, foods that provide assimilable iron, such as lean meat, fish, legumes and whole grains are recommended.

Whereas specific deficiency diseases, such as rickets, scurvy, goitre and xerophthalmia, caused respectively by insufficient vitamins D and C, iodine and vitamin A, can be readily diagnosed, the relation between dietary excess, caloric and nutrient imbalance to chronic disease is complex and diagnostically elusive. Relevant data come from clinical and laboratory measurements of physiological indicators, dietary intake estimates for populations and/or individuals. Epidemiological studies compare these data with specific disease incidence among different communities. Extrapolation of results from laboratory animals to the human condition requires careful consideration since such can be misleading.

Reliable data of nutrient intakes are broadly useful in examining status among large populations, but apparent mathematical correlations between a particular intake and a specific human malady, do not always indicate direct cause and effect. The U.S. Surgeon-General highlights the difference between apparent association and cause as related to diet and chronic disease. Evidence that relates obesity, intake of saturated fats and specific blood cholesterol with risk of CHD is persuasive, as are the benefits to health of complex carbohydrates and dietary fibre.

U.S. NRC Report on Diet and Health

The U.S. NRC Report on Diet and Health is remarkable in its breadth, depth and scholarship. It was the outcome of a study to identify the scientific issues fundamental to nutritional policy related to chronic diseases. The Report points to a lack of consensus on the role of diet in the etiology of chronic diseases, caused in part by incomplete knowledge, in part from disagreement on criteria. Earlier reports had failed to determine how dietary risk factors influence a broad spectrum of atherosclerotic cardiovascular diseases, cancers, diabetes, obesity, osteoporosis, various hepatic and renal disorders (NRC 1989).

The report overlooks little that relates to nutritional adequacy, excess and imbalance in the diets of contemporary Americans. The report begins with a chapter that presents its definitions and methodologies. Later chapters give evidence and critical comments on the major and minor components of diets, on non-nutrients such as tea, coffee and alcohol, and their known and possible associations with chronic disease risks. The report goes on to discuss patterns of diet in relation to such chronic disorders as atherosclerotic, cardiovascular and coronary dysfunctions, hypertension, obesity, various cancers, osteoporosis, diabetes mellitus and dental caries.

The U.S. NRC report assesses data from human and animal studies and presents immense documentation on which its dietary recommendations are based. While explicitly recognising that much remains to be learned, analysis of the evidence available makes possible recommendations to the public about particular associated risks and benefits, and the technological means by which desirable dietary improvement may come about. The recommendations are directed at the general public, and more specifically at individuals whose dietary habits put them at high risk.

The report reaches the general conclusion that diet influences the risk of atherosclerotic cardiovascular diseases, hypertension and some forms of cancer. It arrives at conclusions similar to those of the U.S. Surgeon-General in relation to over-indulgence and obesity. While some disorders may depend in part on environment and genetic inheritance, these have not been indisputably identified in all instances.

The U.S. NRC report concludes that in total amount and chemical nature, lipids in the diet influence various chronic diseases, high fat intakes being associated with obesity, saturated fatty acids and low density lipoprotein (LDL) cholesterol contributing to atherosclerosis and CHD. Omega-6 polyunsaturated fatty acids (PUFAs) lower total and LDL serum cholesterol, omega-3 PUFAs may inhibit certain forms of cancer, reduce plasma triglycerides and raise blood clotting time. Diets high in fruits, vegetables, legumes and whole grain cereals are associated with a reduced incidence of CHD; foods high in vitamin C may protect against stomach cancers. The report regards evidence as too limited to draw firm conclusions relating vitamin E, folic acid and riboflavin to carcinogenesis. It accepts that high fibre intakes may reduce the risk of CHD and colon cancer and provides a critical review of the evidence related to, and uncertainties surrounding ingestion of dietary fibre and reduced risk of CHD and certain forms of cancer.

In adults, obesity results from energy imbalance and, in some degree, from genetic predisposition. Long-term experience indicates that large fluctuations in body weight and excessive restrictions on food intake are undesirable. Blood pressure is positively correlated with daily intake of 6 g or more of salt. Clinical studies demonstrate that hypertension is not invariable reversed by reducing salt intake. People genetically predisposed and those over 55 are at higher risk of hypertension.

A sustained low calcium intake is associated with a high frequency of bone fractures in elderly adults but the potential benefits to alleviation of osteoporosis of dietary calcium in amounts higher than proposed in RDAs are not undeniably established.

In addressing the contentious subject of dietary vitamin and mineral supplements the report states: "The adverse effects of large doses of vitamin A are well documented. There are no documented reports that daily multiple vitamin-mineral supplements, equalling no more than the RDA for a particular nutrient, are either beneficial or harmful for the general population. Potential risks or benefits of the long-term use of small doses of supplements have not been systematically examined."

It concludes with an overall assessment of U.S. dietary habits and offers recommendations to the public at large and to individuals particularly afflicted or at high risk. The report is exceptionally well referenced, each reference being contiguous to the subject under consideration. It is remarkable in its comprehension and scope. The principal recommendations are presented later together with those of the U.S. Surgeon-General.

DIET, DISEASE, OBESITY, OVERWEIGHT

"Obese" is derived from the Latin *Obesus* meaning fat, plump or swollen. In some cultures a corpulent rotundity is favourably regarded. Among certain East Africans very fat ladies are considered highly desirable. According to Shakespeare, Julius Caesar declaimed: "Let me have people about me that are fat" (*Julius Caeser* Act I, Sc. ii).

It is widely accepted by health authorities in the USA that obesity is the most pernicious and pervasive nutritional disorder in America. In 1985, the National Institutes of Health estimated that more than 34 million Americans (over 14% of the population) were significantly overweight (NIH 1985). The terms "obesity" and "overweight" tend to be used synonymously, in many instances correctly so. Yetiv (1987) makes the following distinction: Obesity is characterised by an excess of body fat, more than 20% of body weight among men, over 25% among women.

Overweight is assessed against standards for various states of body architecture. The U.S. Surgeon-General (1988) recommends: "Achieve and maintain a desirable body weight. . . . choose a dietary pattern in which energy (caloric) intake is consistent with energy expenditure. To reduce energy intake, limit consumption of foods relatively high in calories, fats and sugars, and minimize alcohol consumption. Increase energy expenditure through regular and sustained physical activity."

Though available food calories per capita per day have changed little in the USA since the start of the century (*ca.* 3500 kcal), average calories available exceed actual intake, estimated at 2400-2600 kcal for men, 1600-2000 for women. Caloric intake is influenced by age, sex, environment, energy expenditure, body biochemistry, appetite, attitude and self-discipline. Among most Americans, caloric intake peaks during the second decade of life and subsequently falls.

Most notable in terms of health have been the changes over time in the caloric sources. During the century calories from carbohydrates have fallen from 57 to 46%, while calories from fats have risen from 32 to 43%.

Determination of Obesity

Body fat may be estimated by hydrostatic (underwater) weighing, a technique based on Archimedes' Principle, or by measuring skin-fold thickness. A Report of an Expert Group convened by the Health Services and Promotion Branch of (the Department of) Health and Welfare Canada presents guidelines for healthy weights which are not inconsistent with those proposed by other authorities (H&W Canada 1988).

The Canadian Group propose that definition of weight guidelines are consistent with national nutrition program objectives and that the guidelines are intended to provide a scientifically based method of assessment that is simple and practical; that identifies patterns of weight including acceptable low and excess weights that relate to health risks. The weight guidelines are linked to the treatment of obesity, overall health promotion, and are intended to assist health professionals and the general public, defined as "the healthy Canadian adult population 20 to 65 years of age".

The authors review and compare five methods and criteria by which adult weight may be assessed:

1) Weight-Height Tables that relate weight to specific body height and/or frame size. Methods of assessing frame size include measurement of wrist circumference or elbow breadth as set out in the Metropolitan Height and Weight Tables 1959 and 1983, published by the Metropolitan Life Insurance Company.
2) Skinfold Measurements which, using skinfold calipers, estimate subcutaneous fat in the trunk or abdominal regions.
3) Waist:Hip Ratios (WHR) which assess abdominal obesity by measuring the ratio of the waist circumference to the hip circumference. "WHR can be used to

subdivide obese individuals into subgroups which may give a better estimate of their risk of developing health problems".

4) Weight:Height Ratios.

Following a critical comparative evaluation of the following ratios: Weight:Height; Weight:Height squared; and Weight:Height cubed, the Canadian group recommends adoption of the ratio Weight:Height squared. The quotient of Body Weight in kilograms divided by the square of Body Height in metres is defined as the Body Mass Index (BMI) which, the authors state, correlates significantly with body fat as estimated from body density.

From data collected in various health and fitness studies on large samples of the Canadian adult population, the authors classify BMIs into four Zones:

Zone A: BMI below 20 - May be associated with health problems for some low weight people.

Zone B: BMI 20 to 25 - Associated with "good weight for most people" with lowest occurrence of morbidity and mortality.

Zone C: BMI 25 TO 27 - May be associated with ill-health in some adults:sharp increases in morbidity and mortality have been reported at BMIs close to 27.

Zone D: BMI above 27 - Increasing risk to health:BMIs associated with hypertension, hyperlipidemia and CHD.

Table 26 lists BMIs for various heights between 1.5 m (4 ft, 9 in.) and 2.0 m (6 ft, 7 in.) at weights between 65 kg (*ca.* 150 lb) and 115 kg (*ca.* 253 lb).

It is evident from Table 26 that any adult with a weight higher than 95 kg (*ca.* 210 lb) must be of exceptional height (2.0 m = 6 ft, 7 in.) to fall within Zone B, the "good weight" zone.

Obesity and overweight results from too little energetic physical activity and/or over-eating. Increased energy expenditure is inversely related with CHD. Blood pressure levels are significantly positively related to habitual high ingestion of sodium chloride. Epidemiological studies indicate that risk of stroke-related deaths over a wide range of blood pressures is inversely related to potassium intake.

Epidemiological studies consistently illustrate how early mortality and morbidity are causally related to obesity. Obesity aggravates high blood pressure, stroke and CHD. The U.S. Surgeon-General states on average that Americans derive 37% of calories from fat, 13% from saturated fat. Mean cholesterol intake is *ca.* 305 mg/day for women, 440 mg/day for men. Obesity has been associated with gall bladder disease and with cancer of the uterus and breast. More than a quarter of U.S. citizens are believed to be over-weight, a condition prevalent among black women above 45 years of age (Surgeon-General 1989).

Table 26. Body Mass Index (BMI) for various weights and heights.

Wt (kg)	Ht (m)	BMI	Wt (kg)	Ht (m)	BMI	Wt (kg)	Ht (m)	BMI
65	1.5	29	80	1.7	28	100	1.8	31
	1.6	25		1.8	25		1.9	28
	1.7	22.5		1.9	22		2.0	25
	1.8	20		2.0	20			
70	1.5	31	85	1.7	29	105	1.8	32
	1.6	27		1.8	26		1.9	29
	1.7	24		1.9	23.5		2.0	26
	1.8	21		2.0	21			
75	1.6	29	90	1.8	28	110	1.9	30.5
	1.7	26		1.9	25		2.0	27
	1.8	23		2.0	22.5			
	1.9	21						
			95	1.8	29	115	1.9	32
				1.9	26		2.0	29
				2.0	24			

Heights		Weights	
Metres	Ft. in.	kg	lb.
1.5	4'10"	65	143
1.6	5' 3"	75	165
1.7	5" 7"	85	187
1.8	5'11"	95	210
1.9	6' 3"	105	230
2.0	6' 7"	115	253

Lipids and Cholesterol

In 1916, de Langen a Dutch physician working in Java observed that the plasma cholesterol and incidence of coronary heart disease (CHD) was lower among native Indonesians than among Dutch colonists, whose diets were higher in fatty foods. Rosenthal (1934) drew attention to the prevalence of atherosclerotic cardiovascular diseases among people with high intakes of fat and cholesterol. Following WWII, Keys in a series of publications presented data on and a critical assessment of relations between dietary lipid and cholesterol, and CHD and other cardiovascular dysfunctions (Keys 1957, 1975, 1980, 1984). Keys and others demonstrated significant correlations between total serum cholesterol and the mean percentage of calories from saturated fatty acids.

There is persuasive evidence that the quantity and chemical nature of animal and vegetable fats and other lipids in diets can influence the onset of atherosclerotic cardiovascular diseases and, though less well-established, certain forms of cancer. Though high fat intake is associated with obesity, and obesity may contribute to gall bladder diseases, there is no direct conclusive evidence that gall bladder malfunctions are a direct consequence of fats ingested.

Fat available in the U.S. food supply is estimated to have risen from an average of 124 g per capita/day in 1909 to 172 g in 1985, over which period the proportion of animal fat fell from 83 to 58% as Americans consumed less butter and lard, while vegetable lipid from margarine, salad and cooking oils rose from 17 to 42%. Pork and beef are the major sources of animal fat, though, during the past 40 years annual average per capita consumption of beef has dropped from 41 to 36 kg, whole milk from 105 to 55 kg. Over the same time skim milk has risen from 20 to 51, poultry meat from 21 to 32 kg.

Factors that affect lipid intake include the considerable use of oils and fats in fast food restaurants where many foods are fried. Data based upon food supply disappearance do not take sufficient account of losses caused by waste, spoilage, trimming or cooking between the retail outlet and the consumer's table. National food consumption surveys, carried out by the USDA suggest that during the past decade American adults have on average decreased their lipid intake from 41% to 36% of total calories (USDA 1986, 1987). In recent years there is convincing evidence that calories from red meat have declined while poultry and fish consumption have risen (Block et al. 1985).

Clinical, animal and epidemiological studies demonstrate that higher intakes of saturated fatty acids 12-18 carbon atoms in length raise levels of serum total and low density lipoprotein (LDL) cholesterol, these in turn elevating the risk of CHD. A few epidemiological studies on dietary fat and cancer have distinguished between the effects of specific types of fat and indicate that higher intakes of saturated fatty acids and total fat may be associated with increased incidence of colon, prostate and breast cancers (NRC 1989).

Several fatty acids are considered essential to good health. These include linoleic (18 carbons with 2 double bonds), arachidonic (20 C 4 =), eicosapentaenoic (EPA) (20 C 5 =) and docosahexaenoic (DHA) (22 C 6 =). Linoleic and arachidonic belong to the omega-6 group since, counting from the methyl end of the molecule, the first double bond occurs at carbon atom number 6. By the same convention EPA and DHA, present in fish oils, belong in the omega-3 group. Since it is not produced in the human body, linoleic acid must be present in food sources, from whence it is converted to arachidonic acid. There exists a growing body of evidence to suggest that linolenic acid, and its derivatives EPA and DHA are essential in human diets. Linolenic may be converted in body tissue to DHA and EPA, which are also synthesised by phytoplankton and algae and consequently are present in the adipose tissues of finfish and shellfish. Linoleic acid deficiency, determined biochemically by analysis of plasma lipids, may cause several metabolic disturbances, some being associated with dry and scaly skin and hair loss. Linoleic acid intake at levels of 1.0-2.0% of total dietary calories is generally sufficient to prevent biochemical and clinical symptoms of deficiency. As yet there is insufficient scientific data from which to prescribe RDAs for omega-3 and omega-6 polyunsaturated fatty acids.

Kromhout et al. (1985) were intrigued by the low death rate from CHD among Greenland Esquimaux whose diet is rich in fish. Patterns of fish consumption and death from CHD among 852 middle-aged Dutchmen were followed over a 20-year period. An inverse dose-response relation was observed between fish consumption from 1960 and death related to CHD from 1960 to 1980. Mortality related to CHD was 50% lower among men who ate at least 30 g of fish per day than among non-fish eaters. It was concluded that fish once or twice per week can be preventive of CHD.

Phillipson et al. (1985) studied fish oils in the diets of 20 hypertriglyceridemic patients, the diets differing primarily in fatty acid composition and fat content. The control was typical of a low-fat therapeutic diet, a fish-oil diet contained omega-3 fatty acids, a vegetable-oil diet was rich in omega-6 linoleic acid. The fish oil diet led to a decrease in both plasma cholesterol and triglyceride, and a marked reduction in Very-low-density-lipoproteins (VLDL). The diet high in vegetable oil was less effective. It was concluded that fish and fish oils can be beneficial in treating hypertriglyceridemia.

Glomset (1985) studied omega-3 and omega-6 polyunsaturated fatty acids (PUFAs) and serum cholesterol, concluding that different PUFAs produce different metabolic effects. Glomset discusses omega-3 PUFAs in human health and disease, and offers several alternative mechanisms by which beneficial actions may take place and presses for more research to elucidate the underlying principles.

Tso and Weidman (1985) report that lipids provide at least 40% of the daily caloric intake throughout most of the western world, lipid intake ranging between 60 and 100 g per capita per day, most in the form of triglycerides, the long chain oleic and palmitic being in highest proportion. In human milk and in infant formulas 40-50% of calories are present as fat.

Beneficial biological attributes of fish oils, of eicosapentaenoic (EPA) and docosahexaenoic (DHA) have been reviewed by Carroll (1986), Herold and Kinsella (1986), and Glomset (1985). EPA is found in the lipids of fish but not in the fat of terrestrial animals. Clinical and animal studies indicate that omega-6 PUFAs lower both total serum cholesterol and LDL cholesterol, with some increase in high density lipoprotein (HDL). Fish oils, rich in omega-3 PUFAs may reduce plasma triglyceride levels and accelerate blood clotting. Beneficial effects of fish oils in prevention of thrombosis and atherosclerosis are attributed to EPA. Prostaglandins, helpful to arterial function, are synthesised from EPA.

Diets high in foods of plant origin:fruit, vegetables, legumes, nuts and whole grain cereals are believed to reduce risk of CHD, cancer of the lung, oesophagus and stomach. Absolute vegetarians and lacto-ovo- vegetarians display serum levels comparatively low in triglycerides, total and LDL cholesterol. Epidemiological studies indicate that intake of foods rich in carotenoids are inversely associated with lung cancer risk; and that foods high in vitamin C may offer protection against stomach cancer induced by nitrosamines. Clinical and epidemiological research results suggest that diets high in fibre may reduce risk of CHD, colon cancer and hypertension. Inter-country correlations indicate diets high in red meat are positively associated with atherosclerotic CHD and cancers of breast and colon.

Though slight elevation of serum cholesterol has been attributed to drinking coffee, epidemiological results have not persuasively shown coffee consumption to be a contributing cause of CHD or cancer, such evidence as has been offered being weak and inconsistent (NRC 1989).

Unsaturated fatty acids that occur naturally exist mainly in the *cis* form. During conversion of vegetable oils to plastic fats by catalytic hydrogenation, some of the fatty acids change to *trans* isomers. Though small quantities of *trans* isomers are found naturally in dairy fats, what may be the nutritional consequence of higher intakes of *trans* isomers from hydrogenated oils is at present unclear but is cause for concern among some observers (Yetiv 1987; Scrimshaw 1992).

Diet and Psychological Disorders

Yetiv (1987) reviews the effects of nutrients on psychological functions, in particular, precursors of specific neurotransmitters that occur in foods. Laboratory studies with animals indicate that choline and lecithin can raise levels of the neurotransmitter acetylcholine. In a critical analytical review of cholesterol, lipids and cardio-vascular disease Yetiv (1987) cites over 160 references, most being from medical literature published between 1974 and 1984.

General Comments on Dietary Lipids

Scientific interest in HDL was aroused when Miller and Miller (1975) and Miller et al. (1977) showed evidence that HDL may protect against CHD. Miller and Miller (1975) proposed that a reduction in plasma HDL concentration may accelerate the development of atherosclerosis and ischaemic heart disease by impairing the clearance of cholesterol from arterial walls.

Experimental and clinical evidence strongly supports the hypothesis that total cholesterol and low density lipoprotein (LDL) are causally implicated in CHD whereas high density lipoprotein (HDL) is inversely correlated and presumed protective. The U.S. Surgeon-General, U.S. NRC and others recommend that total blood cholesterol and HDL be monitored in all adults and more frequently in those whose total cholesterol is found to be over 250 mg/dL (Surgeon-General 1988; NRC 1989; FNB/NRC 1989).

ADEQUACY, EXCESS AND NUTRIENT DEFICIENCIES

Protein and Amino Acids

The average protein available in the U.S. food supply approximates to 100 g/capita/day. Since the start of this century, protein from animal sources has risen from *ca.* 50 to 68% of total available protein. The U.S. Department of Agriculture, from Nationwide Food Consumption Surveys, estimates a mean protein intake of 74 g/capita/day which furnishes a little over 16% of caloric intake (USDA 1984-1989).

Though diets high in grains, fruits and vegetables are less rich in protein than those high in livestock products, average daily protein intakes for American women (60 g) and men (90 g) are above U.S. RDAs of 44 g for women and 56 g for men. In the USA, meat, poultry and fish provide about 50%, dairy products 18%, eggs 4%, legumes 3%, cereals 18%, fruits and vegetables 7% of average total dietary protein.

Because of a close relation between animal protein and saturated fat intake, their independent effects on health are difficult to isolate. Some epidemiological studies have compared the nutritional status of vegetarians with meat eaters. U.S. NRCs review of diet and health states: "the total protein intake of vegetarians is not much lower than that of omnivores; however, the life styles of vegetarians may differ from those of omnivores in ways that may confound the association between vegetable or animal protein intake and health " (NRC 1989).

In the international literature much of the data on protein and health is derived from studies of chronic protein malnutrition, many of the subjects having relatively low blood pressure. Several studies reported by NRC (1989) found lower systolic and diastolic pressures among lacto-ovo-vegetarians than omnivores.

Tryptophan is converted to the neurotransmitter serotonin believed to be involved in the control of pituitary hormone secretion, sleep patterns and perceived pain. Fernstrom and Wurtman (1971, 1972) report that foods low in protein, high in carbohydrate, tend to raise the level of tryptophan in the brain, whereas foods high in protein give rise to the opposite effect. Administration of tyrosine, a precursor of dopamine, is reported as effective in relieving depression (Yetiv 1987).

Vitamins

Yetiv presents evidence to suggest that certain vitamin deficiencies may cause neuropsychiatric disorders which respond to supplements of the appropriate vitamin. Deficiency syndromes related to impaired mental functions have been related to inadequate niacin, thiamin, ascorbic acid and vitamin B12. Whether moderate deficiencies of the vitamins are accompanied by cognitive dysfunctions, sometimes apparent among the elderly, is not clear (Yetiv 1987).

The U.S. Surgeon-General estimates that 40% of Americans buy vitamin and mineral supplements at an annual cost of $2.7 billion. No demonstrable advantages are reaped by normally healthy people who consume vitamins and minerals in exceptional quantities (Surgeon-General 1988).

Adult Americans, on average, consume between 4 and 6 g of sodium; the daily intake recommended by the National Research Council is 1.1-3.3 g. Studies of animal protein, sugar and coffee consumption show variable associations with blood lipid levels, the relations being too uncertain to draw any specific conclusions relevant to ill-health (NRC 1989).

Epidemiological studies suggest that dark green vegetables of the Cruciferae family may protect against cancer of the alimentary tract and bladder, and that certain carotenoids, such as precursors of Vitamin A, may protect against epithelial cell cancer in oral cavities, bladder and lung.

FADS, FANCIES, AND UNCONVENTIONAL DIETS

In his Preface to *The Doctor's Dilemma* G.B. Shaw counsels: "Do not try to live forever. You will not succeed." Despite this axiomatic advice, there is a vast literature variously offering realisation of the alchemists' hope of an *Elixir vitae*.

A wide-ranging, scientifically critical, well-referenced review of various reputable and fanciful publications is presented by Yetiv (1987). The author covers: reducing body weight; herbal remedies; "natural" foods; vegetarian and liquid protein diets; vitamin, mineral and other nutrient supplements; effects of salt, saturated fat and cholesterol; diet and psychological behaviour; allergies and intolerances; eating disorders; some specific "Fad Diets".

He describes as "a dietary fad with no apparent benefit" the use of "starch blockers" some 200 of which were offered for sale in health food stores during the 1980's. The active principle was phaseolamin extracted from raw Phaseolus beans which inhibits amylase and thereby, according to its promoters, would block the digestion and absorption of ingested carbohydrate. Controlled laboratory studies that refuted the starch-blocking hypothesis are described (Bo-Linn 1982; Garrow 1983). At first, being marketed as dietary aids they were considered outside U.S. FDA regulatory control. Later, in 1982, FDA embraced them within their jurisdiction by classing them as drugs. In the meantime a million or more "starch-blocker" tablets were swallowed every day.

Among advocates of "natural foods" it is believed that all organic materials of natural origin may be presumed safe and wholesome, but whatever emerges from a biochemical factory is highly suspect. A trend towards minimally processed foods, including vegetables packaged under controlled atmosphere, is evident in the USA and Europe.

Yetiv (1987) refers to potent drugs extracted from plants which may first have appeared in herbal remedies and which, dependent on dosage level, may be remedial or fatally toxic. Yetiv describes particular herbal and "natural" remedies that have proved beneficial, others that have caused distress and discomfort. He emphasises that the descriptor "herbal" guarantees neither safety nor effectiveness; only rigorous scientific experiments can determine what can be a remedy for what ailments, under what conditions. Yetiv infers that unbiased scientific experiments are generally lacking in "the herbal and alternative health care arena."

Vegetarian diets fall into three broad categories:

Vegan - strict vegetarian: devoid of all foods of animal origin

Ovolactovegetarian: free from animal flesh; eggs and milk permitted (In some cases only unfertilized eggs are acceptable.)

Semi-vegetarian: various regimens, some permitting fish but not meat from terrestrial animals; some fish and poultry meat but not red meat.

Yetiv (1987) outlines evidence that vegetarian diets impose a lower risk of cardiovascular disease than non-vegetarian. Being relatively high in fibre and complex carbohydrates, vegetarian diets can be of benefit to diabetics and reduce the severity of diverticulosis and other colonic ailments.

Animal proteins are nutritionally desirable in that they contain all essential amino acids in proper proportions. Vegetable proteins are incomplete, cereals being generally deficient in lysine and threonine, legumes and oilseeds in methionine and cystine. Taking advantage of the amino acid complementarity of cereal and legume proteins, vegan diets can provide nutritionally adequate protein if suitably formulated and prepared (Hulse 1991).

Strict Vegans may be at risk of deficiencies in calcium, vitamin B12 and vitamin D. Vitamin B12 deficiency among members of a Rastafarian cult, led to neurological and haematological disorders which were remedied by supplementation of the diet with vitamin B12 (Yetiv 1987).

Yetiv presents an edifying critique of a massive, widely accepted publication by Durk, Pearson and Sandy Shaw: "Life Extension: A Practical Scientific Approach" which proposes that various morbidities and malignancies can be prevented or contained by megadoses of particular nutrient supplements, with a consequent extension of human life span by several decades. Yetiv questions the scientific reliability of the advice given, it being difficult to determine from the "Life Extension" text which of the propositions derive from sound scientific experiment and which are anecdotal, hearsay or imaginary.

There appears to be an inexhaustible demand for diet books that offer a painless and novel prescription for weight loss or some other physiological correction. In spite of abundant scientifically sound and sensible information and advice on diet and health from government agencies, addictions to quick-fix faddish and fanciful diets appear as endemic among many Americans. The advice given by the U.S. Surgeon-General, the National Academy of Sciences, the Food and Drug

Administration and many others: to maintain a low intake of fats in general, of saturated fats in particular, and in cases of obesity to decrease caloric intake and increase physical activity, would seem meaningful and applicable even to those of modest intelligence and education.

Resort to such extremes as drug and surgical therapy should be viewed with extreme caution. Drugs may act as appetite depressants or in other ways not fully understood. Surgical procedures include intestinal and gastric by-pass, and suction lipectomy, the latter carried out by plastic surgeons who use a cannula probe to extract fat from selected adipose tissue (Yetiv 1987).

Unconventional Diets

Popular diets of various compositions are reviewed by Yetiv (1987) and Truswell (1982). Yetiv critically assesses liquid protein diets, popular during the 1970's, which resulted in fatal consequences for some adherents; the Beverly Hills Diet which permits only fruit for the first 10 days; the liquid formula Cambridge Diet, said to provide 33 g of protein, 44 g of carbohydrate in a 300 kcal formula; the F-Plan diet, high in fibre and complex carbohydrate; the General Foods Setpoint Diet based on the questionable proposition that each individual's metabolism adjusts to a particular body weight or "setpoint". The setpoint level is not altered by dieting alone and regular exercise must accompany low calorie intake.

Truswell (1982) offers interesting observations on popular diets for weight reduction. Some advise carbohydrate restriction while allowing foods high in fat and protein ad libitum. One described as "a nutrition break-through" based on meat and millet claims to be non-ketogenic and adequate in fibre. Diets high in protein are commonly featured among the panaceas proposed. One such recommends lean meats accompanied by eight glasses of water plus multivitamin tablets. The Scarsdale Medical Diet recommends 55% of energy from protein, 22% from fat with a menu cycle with unlimited amounts of specified meat and fish, plus salads, fruits and very little bread.

Truswell comments that though foods high in protein may appear as gustatorily satisfying, few if any scientifically controlled trials that determine the satiety value of different foods have been reported. Clinics that purport to specialise in weight reduction are a growth industry in the USA. Fees in excess of $250 per day are charged for the privilege of dietary regimens that could be equally well adopted less expensively at home. One such diet is high in whole grain cereals, vegetables and fruits, low in sugar and fat, free from salt, alcohol and caffeine.

Some resort to such oddities as, successively, apples, cheese, chicken, salads each for two days. A Beverly Hills Diet provides only fruit for 10 days, bread, salad and meat being progressively added later. Truswell quotes the *Journal of the American Medical Association*'s description of the Beverly Hills Diet: "The latest and perhaps the worst entry in the diet fad Derby. The Diet's major tenets fly in the face of all established medical knowledge about nutrition." Nonetheless, some 800,000 copies of the hard back edition were sold in one year. A diet designed for CHD victims partially replaces saturated fats with polyunsaturated oils. The F-Plan Diet from Britain offers high fibre from cereals and fruit, low fat recipes high in skim milk and legumes.

Most unconventional weight-reducing diets seem to originate in North America where, despite the abundant reliable advice freely available from health ministries and other sources, for some obese folk, the only means to body weight control seems to be by residence in expensive clinics or recourse to curious diets.

INTOLERANCIES AND ALLERGIES

In the popular press and commercial advertising "food allergies" have been held responsible for various indispositions, including dermal and respiratory irritations, nervousness, fatigue and depression. It is sometimes difficult to distinguish scientifically verified cases from instances based on anecdotal or other inconclusive evidence. Immediate and severe reactions to particular foods may be quickly and reliably recognised and diagnosed as attributable to specific allergies. In the case of milder reactions, it is more difficult to define specific cause and effect, and to distinguish between symptoms that are more probably psychological than physiological or biochemical in nature. Several observers suggest that psychopathology may be misleading and that double-blind testing is more dependable in diagnosing specific allergies (Yetiv 1987; Pearson et al. 1983).

Yetiv (1987) discusses food intolerances and allergic responses to various food additives including sulphites used as preservatives and sometimes associated with asthma in susceptible consumers; monosodium glutamate extensively used in Oriental cooking and in processed meat products, which has been reported to cause mild ataxia, abdominal pain and asthmatic reactions. Check (1983) contends that in some instances migraine may be induced by food to which sufferers are sensitive.

Pearson et al. (1983) define "food hypersensitivity as an organic adverse reaction to a food that is not due to its inherent toxic properties, and food allergy as hypersensitivity associated with evidence of an immunological reaction". The authors sought evidence of food hypersensitivity by the use of exclusion diets and double-blind provocation tests in 23 patients who had attributed a wide variety of symptoms to food allergy.

The authors could not detect hypersensitivity to ingested substances in 19 patients; hypersensitivity was confirmed in only four cases, each of whom displayed typical atopic symptoms. None of the four showed psychologically related symptoms, but a high incidence of psychiatric disorder was found among patients whose beliefs of allergic affliction could not be experimentally confirmed, a diagnosis that generated a somewhat hostile response. The authors conclude that, to offset possible psychogenic reactions, double-blind test methods are the most reliable means of diagnosing biochemical hypersensitivity to foods.

Truswell (1985) describes food sensitivity as a subject "at the interface between scientific immunology, food technology, and quackery". Truswell presents definitions compatible with what are presented in a report of the Royal College of Physicians (1984). These include:

Food sensitivity: Reproducible unpleasant reaction to a specific food or ingredient which one person gets but others do not.
Food allergy: Evidence of abnormal immunological reaction to food.
Food intolerance: Various mechanisms other than allergy.
Food aversion: Psychological and psychosomatic reactions.

Truswell discusses diagnostic methods and dietary manipulation but cautions: "There are no straightforward diagnostic tests for food sensitivity comparable to the electrocardiogram for coronary disease". Various responses common to particular sensitivities are described: urticaria and angio-oedema, rhinitis and asthma, eczema, migraine and hyperactivity in children, together with the foods and other ingested substances that may cause onset.

Check (1983) discusses the controversies that impinge on allergies: uncertainties in symptoms from food allergies and other intolerances; and comments on various provocative and cytotoxic diagnostic methods, and the contrasting views on their respective merits.

CONCLUSIONS AND RECOMMENDATIONS ON DIETARY IMBALANCE

The U.S. Surgeon-General's Report

The U.S. Surgeon-General's report adds weight to the recommendations contained in "Dietary Guidelines for Americans" issued by the U.S. Departments of Agriculture, Health and Human Services. Americans are advised: (1) to eat a variety of foods, (2) maintain desirable weight, (3) avoid excessive fatty foods, saturated fatty acids and cholesterol, (4) choose foods high in complex carbohydrates and fibre, low in sugar and sodium (Surgeon-General 1988).

The U.S. Surgeon-General's report carries serious policy implications for the public at large and people of particular conditions. Governments and educational authorities bear responsibility for promulgating sensible advice on diet and health: on what constitutes a desirable dietary balance; on caloric intake relevant to weight, height and activity. Health professionals and medical practitioners seem often less well informed about nutritional principles and practices than their responsibilities demand. Prophylaxis, by adoption of sensible eating habits, is more economic and efficacious than clinical measures required to reverse and remedy diseases and discomforts from dietary excess and imbalance. Consumers should be educated to understand the nutritional implications of what is set out in food labels. There is reason to believe that relatively few even read let alone understand labelled contents.

The U.S. Surgeon-General looks to food industries to provide processed foods so composed and convenient to use as to promote healthy eating habits. Nonetheless, food industries provide what consumers are persuaded to buy and eat, which may not always be what is best for their health.

Research Priorities

Recognising the many and diverse lacunae in exact nutritional knowledge, the U.S. Surgeon-General recommends several priorities for research and surveillance:

1. Relations between specific dietary factors and the etiology and prevention of chronic diseases possibly related to food;
2. Dietary patterns for children conducive to protection of present and future good health;
3. Effects of maternal nutrition on the health of developing foetus;
4. Nutrient and energy requirements of older adults and whether they are desirably different from those of younger people;
5. Identification of biochemical markers of dietary intake to monitor effects of dietary interventions;
6. Educational programs that encourage wise food choices;
7. Food surveillance systems to monitor population-specific and state-specific trends in risk factors.

Research on obesity proposed includes:

- Desirable weight-to-height for different age groups;
- Specific health risks associated with degrees of over-weight;
- Precise methods to measure total body fat and fat distribution;
- Relation of metabolic and genetic factors;
- Influence of genetic aberrations;
- Relations of diet, physical exercise, body weight and weight loss on metabolism and thermogenesis;
- Influence of psychological and sociological factors;
- Health consequences of repeated cycles of weight gain and loss;
- Intervention strategies for weight management.

The U.S. Surgeon-General discusses osteoporosis in relation to calcium, phosphate, vitamin D, protein and caloric intake, and proposes research on these and other dietary habits on bone mineralisation as people advance in age. Research is needed on possible adverse effects of high-dose supplements of calcium, calcium sources higher in bioavailability; vitamin D, its metabolites and calcium on the development of peak bone mass and prevention of bone deterioration.

The report discusses diet and dental disease, kidney malfunctions, dietary fat, fibre and alcohol related to gastro-intestinal maladies, their prevention and alleviation.

Research is proposed on interactions between diet, infection and immunity, including studies of mechanisms by which malnutrition may depress particular components of the immune system; specific elements and micro-nutrients in food that may influence adult immune systems, or in breast milk that may affect the health of infants by transmitting immunities, toxins or allergens.

Many other research subjects related to diet and health proposed include: Prevalence and incidence of food-borne infections, intolerances and immunological reactions to specific food components; susceptibility to infection in relation to nutritional status; behavioural and social factors that influence dietary habits and a greater or lesser risk of ill-health. Research is recommended on attitudes and behaviour that lead to obesity, anorexia nervosa and bulimia, and their prevalence. The S-G prescribes studies on nutrient intake and weight gain during pregnancy; the effects of maternal diets and habits on foetal health; methods of teaching parents sound nutritional habits.

For the elderly, attention should be given to possible desirable changes in nutrient and caloric intake and their influence together with life style and environment on longevity and physiological states, covering people in institutions and those who live alone.

Though the U.S. Surgeon-General's report is based upon experience in the USA and is addressed primarily to Americans, the observations and recommendations are relevant to other communities including those in developing countries where chronic consequences of excessive eating and nutritional imbalance are readily evident (Surgeon-General 1988).

U.S. NRC Report on Diet and Health

The principal recommendations of the U.S. NRC report are:

1. Reduce total fat intake to no more than 30% of calories. Reduce saturated fatty acid intake to less than 10% of calories and or cholesterol to below 300 mg/day.
2. Each day eat five or more servings of a combination of vegetables and fruits, especially green and yellow vegetables and citrus fruits.

3. Increase intake of starches and other complex carbohydrates.
4. Maintain protein intake at moderate levels, equivalent to less than 1.6 g/kg of body weight for adults.
5. Balance food intake and physical activity to maintain appropriate body weight.
6. Limit total daily ingestion of salt to no more than 6 g.
7. Avoid dietary supplements in excess of the prescribed RDA.

Research Priorities

The report also make recommendations for research priorities many of which are similar to those proposed by the U.S. Surgeon-General, including:

1. Research to identify foods and dietary components that relate to risk of chronic diseases and to elucidate the mechanisms of action.
2. Improved methodologies by which to collect and assess data that relate to disease risk, to quantify adverse and beneficial dietary effects and determine optimal ranges of macro and micro constituents.
3. Basic research on molecular and cellular biochemistry with more effort to apply scientific knowledge to public health programs.

The extent of agreement between the recommendations of the U.S. Surgeon-General and the U.S. NRC committee should be persuasive to all Americans and in varying degrees to other comparatively affluent societies. Nutritional disorders and deficiencies among the poorer nations, indeed among many impoverished and disadvantaged Americans, differ markedly in kind and magnitude from those of over-indulgent affluent people, and in many instances are less susceptible to correction and amelioration through voluntary choice and change of habit. Nonetheless, the two U.S. reports provide a wealth of responsible, reputable and reliable advice and information for all who are involved and seriously interested in human nutrition and food security.

CARBOHYDRATE AND DIETARY FIBRE

Complex Carbohydrate

Consistent with those of other nations, the Government of Canada's "Nutrition Recommendations" suggest that in Canadian diets no more than 30% of energy should be derived from fat, and not less than 55% should come from a variety of carbohydrate sources.

Stephen (1994) has reviewed evidence to indicate possible benefits from raising the proportion of starch in the diet. Stephen refers to surveys which suggest that reduced fat intake among North Americans has been compensated more by higher protein than carbohydrate. For many years starch was widely perceived more as a filler and a fattener than as a nutrient with uniquely useful properties.

Stephen comments on patterns of intestinal digestion and sources of different types of starch including: starch that is rapidly digestible, slowly digestible, and resistant to digestion. The resistant forms Stephen sub-classifies into starch that is physically inaccessible, present in partly milled grains and oilseeds; resistant starch granules present in raw potato, bananas and plantains; retrograded starch present in various cooked and baked foods.

Stephen refers to research indicating that *ca.* 10% of ingested starch may pass through the small intestine to be subsequently fermented by microflora in the colon. Stephen's paper suggests that fermented starch yields a higher proportion of short-chain fatty acid butyrates than does similarly fermented dietary fibre. Butyrate is a preferred energy source for the colonic mucosal cells and there is some evidence from *in vitro* experiments to suggest certain anti-carcinogenic properties (Stephen 1994).

Dietary Fibre

Since the early 1970's nutritionists' interests have been much drawn to dietary fibre. A search of the U.S. National Agricultural Library literature data base (CAIN/Agricola) revealed that during the period 1970-73 there were 12 accessioned papers on dietary fibre. Between 1974 and 1977 the number rose to 155, from 1978 to 1981 to 287. Fascination with fibre has not yet abated.

Dietary fibre is a complex material, highly variable in composition among sources. The scope of the designation "dietary fibre" has been redefined several times (Trowell 1972; Trowell et al. 1976, 1978; Southgate 1982; LSRO 1987; Selveldran 1984). Trowell (1972) defined dietary fibre as "the remnants of the plant cell wall that are not hydrolised by the alimentary enzymes of man" which would include cellulose, hemicellulose, pectin and lignin. Later, Trowell et al. (1976) described dietary fibre as a complex mixture, not a distinct chemical entity, the composition being variable among different food types. They expanded the definition to "plant polysaccharides and lignin which are resistant to hydrolysis by the digestive enzymes of man", which include such indigestible plant materials, not necessarily found as components of cell walls, as polysaccharides from guar and locust bean, mucilages, alginates and carageen extracted from aquatic plants. The term "edible fibre" is given a further distinction by Trowell et al. (1978).

The Life Sciences Research Office of the Federation of American Societies for Experimental Biology later defined dietary fibre as: "the endogenous components of plant materials in the diet which are resistant to digestion by enzymes produced by humans" (LSRO 1987). This definition excludes various substances accepted as dietary fibre by some other observers.

Since "dietary fibre" is a general and not a specific designation, the effects and consequences of fibre ingestion are many and varied dependent upon which of the many kinds is under consideration. Overall, ingestion time is slower than for substances more easily digested; some fibres delay gastric emptying, some stimulate faecal excretion of bile acids, some seem to enhance glucose tolerance (LSRO 1987). Dietary fibre may affect colonic microflora and functions, by raising the bacterial mass but not altering microfloral composition.

Quantification of total dietary fibre is difficult since for many years food composition data presented only crude fibre determined by successive treatment with sulphuric acid and sodium hydroxide of defatted plant materials. The method was subject to significant error and results are generally at best approximates. Southgate (1982) and Selvendran (1984) present helpful reviews of sources, composition and methods of analysis for dietary fibre from different sources.

The popular press has been enthusiastic in advocating that foods high in fibre may reduce risk of colon cancer. Journalists among others with limited scientific knowledge and understanding may not always appreciate that fibre exists in many forms and compositions, and that physiological response is uncertain among different sources of fibre, which can vary in such properties as water-binding capacity, viscosity, susceptibility to, rate and nature of bacterial fermentation, and the time taken to pass through human intestines.

DIETARY INADEQUACY AND IMBALANCE

Though incidences of chronic and transitory malnutrition are found among all nations, particularly among poor and mentally maladjusted people, the main concerns of governments in the USA and other OECD nations, are with chronic diseases and imbalances caused by excessive indulgence. The wealthiest nations represent 16% of global population and enjoy 82% of world GNP with an average per capita income 5 times that of the poorest who make up 61% of population with only 6% of global GNP. It is among these poorest nations that population growth is most rapid and food security is at highest risk. Table 3 shows the pattern of wealth and demographic distribution between 1967 and 1990.

The worsening plight of the poorest nations, particularly those in the drought-prone regions of Africa, is the subject of many recent publications. The fulsome publicity lavished on the so-called Green Revolution may have inspired a complacent belief that science and technological innovation will satisfy all future demand for food no matter how large the planet's population. Without denigration of the admirable progress realised in food crop improvement, undue complacency overlooks the fact that during the 1980's in 75 developing countries food production per capita declined. Pinstrup-Andersen (1994) points out that apart from Africa, 80% of food crop production gains in the 1980's resulted from higher yields (weight of grain harvested per hectare of land) and that the area of good quality arable land is falling in many regions so that future food security for many millions will depend upon achieving even higher crop yields.

The FAO/WHO Plan of Action for Nutrition states "despite considerable progress in recent decades, the world still falls far short of the goal of adequate food and nutrition for all people". "Over 780 million do not have enough food to meet basic daily needs for energy and protein." (FAO/WHO 1992).

Nutritional Deficiencies Among the Poor

More than 2 billion people subsist on diets deficient in essential vitamins and minerals, deficiencies that cause such disabilities as blindness and mental retardation. Many more suffer from lack of access to sources of safe, clean water (FAO/WHO 1992).

In 1992, the United Nations Subcommittee on Nutrition published its report on the state of world nutrition, a report that illustrates how protein-energy and micro-nutrient malnutrition continue to afflict many millions in the Third World (ACC/SCN 1992). Roughly 20% of the planet's people is undernourished, one-third of developing country children suffer retarded growth, more than 40% of women are underweight and/or anaemic. The report states that "at least one billion people worldwide are probably affected by nutritional deficiencies."

Table 27 shows incidence of underweight among preschool children and that while the overall percentage has fallen during the past 25 years, absolute numbers have increased globally, in sub-Saharan Africa and in south Asia. Table 28 presents an overview of malnutrition in developing countries showing that from the mid-1970's to 1990 both the percentage and absolute numbers of people with energy intake below what is desirable declined. Nonetheless, close to 800 million, about 20% of all developing people, have access to too little food to maintain healthy active lives. Table 28 shows the numbers and proportions of people who suffer from iron, iodine and vitamin A deficiencies. Close to 140 million children suffer vitamin A deficiency so severe as to cause xerophthalmia. Scurvy, pellagra and beri-beri, caused respectively by too little vitamin C, niacin and thiamin, which have not been in evidence for many years now afflict an inestimable number of displaced people in refugee camps with access to food too little in quantity, variety and nutritional quality.

Table 27. Incidence of underweight preschool children.

	Percent underweight		Numbers underweight (millions)	
	1975	1990	1975	1990
Sub-Saharan Africa	31.4	19.9	18.5	28.2
N East/N Africa	19.8	13.4	5.2	4.8
S Asia	67.7	58.5	90.6	101.2
SE Asia	43.6	31.3	24.3	19.9
Central America/ Caribbean	19.3	15.4	3.4	3.0
S America	15.7	7.7	4.8	2.8
Global Totals	41.6	34.3	168.0	184.0

Changing patterns of available food supply are illustrated in Tables 12, 13, 14, 15, and 16 with data assembled from various FAO publications (FAO 1992a, 1992b, 1993). Table 16 shows how in all regions other than sub-Saharan Africa total food harvested per capita has improved. Despite an average overall increase of close to 3% per annum, in the Near East, south Asia and Latin America production is barely keeping ahead of population growth. Table 14 lists production indices, compares crop and livestock production in 1985 and 1991 with the end of the 1970's. The performance of the eastern European countries who formed the USSR gives cause for concern in that production per capita has fallen since 1985, added to which systems of marketing and distribution between farms and urban populations appear woefully inefficient. Table 15 gives cereal production data and self-sufficiency among developing regions in 1970 and what is forecast until 2020. Collectively, only the nations of south and east Asia are expected to remain near self-sufficient in cereal crops, though Asian arable land is down to barely 0.2 ha per capita.

Table 28. Malnutrition in developing countries.

	Percent		Numbers (millions)	
	1974-76	1988-90	1974-76	1988-90
GENERAL MALNUTRITION				
Too low energy intake				
Total LDC population	33	20	976	786
Children under 5 yr	42	34	168	184
	%		Millions	
MICRONUTRIENT DEFICIENCIES				
Iron deficiency anaemia				
Women 15-49 yr	42		370	
Children under 5 yr	51		-	
Iodine deficiency (endemic goitre)				
All ages	-		200-300	
Vitamin A deficiency (xerophthalmia)				
Children	-		40	

SOURCE: ACC/SCN (1992); Scrimshaw (1993).

The International Food Policy Research Institute (IFPRI), early in 1994, was host to a meeting which discussed future food supply, the environment and population growth. It stated that global population is rising faster than ever before and that policies enacted before 2000 will decide the course of events well into the next century. There was general agreement that by 2010 food will be theoretically sufficient to meet global needs but several regions will be in imbalance. In Asia, food supply per capita is expected to remain relatively constant; among sub-Saharan African nations deficits will likely persist. Economic and environmental consequences of meeting production requirements cannot be predicted. Most production increase must come from higher yields since barely 25% of rising food demand in the Third World can be met by expansion of cultivated land. An IFPRI Conference in 1995 re-emphasised these predictions (IFPRI 1995b).

Rapidly rising demand for water by industries and expanding urban communities will be in competition with agricultural needs. The IFPRI meetings expressed concern over the declining support for agricultural research since 1980, particularly decreased investment in irrigation management, technology improvement and support for national agricultural systems in developing countries.

In a paper that presents the challenge of global malnutrition to food industries, Scrimshaw (1993) writes: "Despite the public attention paid to famine and starvation, most of the hunger that is damaging the survival, development, and welfare of under-privileged populations is unrecognised. This is because even where there are clinical symptoms, they are not associated with food intake. It is this hidden hunger that is responsible for most of the excessive mortality and ill-health afflicting developing country populations, and for the permanent impairment of physical capacity and cognitive performance." Scrimshaw illustrates how when the immediate cause of death is diagnosed as diarrhoea, pneumonia, measles or other infections, such deaths would rarely have occurred in well-nourished infants. Clinically demonstrable marasmus, kwashiorkor, beri-beri, pellagra and scurvy, at one time highly prevalent, have largely disappeared except among refugees.

Scrimshaw discusses three micro-nutrient deficiencies critical to child morbidity and mortality: iron, vitamin A and iodine. Scrimshaw proposes iron deficiency as the most prevalent nutritional problem; iron deficiency without clinically evident anaemia can exacerbate infection, morbidity and mortality. Body functions first affected by iron deficiency are brain enzymes and neurotransmitters that affect cognitive ability and behaviour. Next to suffer are cells of the immune system, a number of anti-microbial mechanisms within the neutrophile being adversely affected. Among child-bearing women, mortality, prenatal and perinatal infant loss, and premature delivery can result from iron deficiency.

Scrimshaw contends that over 40% of Asians and Africans, and in some Indian states 60 to 80% of pregnant women are anaemic. To alleviate anaemia, Scrimshaw recommends double fortification of salt with iodine and iron. Fruits and green vegetables would provide vitamin C to raise iron absorption and reduce vitamin A deficiency. Scrimshaw describes the physiological functions attributable to vitamin A, pointing out that night blindness is an early indicator of deficiency which, when severe, can cause lesions leading to permanent blindness. He estimates that some 40 million children suffer from too little vitamin A, and as a result, each year 350,000 infants become blind. Scrimshaw gives evidence of reduced infant mortality in consequence of supplementary vitamin A administered over a 12-month period.

Scrimshaw (1993) comments that until recently iodine deficiency was associated with cretinism and endemic goitre:a compensatory swelling of the thyroid gland, conditions manifested in children as a result of severe iodine deficiency during pregnancy. The neurological consequence may be more subtle. In any event, extensive iodine deficiency is evident in at least 90 countries, with close to one billion people at risk. Endemic goitre is believed to afflict over 200 million, at least 10% of whom experience some degree of mental retardation or neurological disorders. Iodine deficiency is relatively easily prevented or remedied by fortifying common salt with potassium iodide or iodate.

Scrimshaw comments on undernutrition and inadequate energy intake, to survive which people either reduce physical activity and/or lose body mass, the social and economic eventualities of which are extreme though difficult to quantify. Scrimshaw estimates that close to 63% of the Ethiopian population and 50% of residents in 10

Indian States must be regarded as seriously undernourished.

Scrimshaw reviews the criteria for protein-calorie malnutrition in children, most evident being growth retardation. The first UN report on world nutrition in 1987 adopted as "undernourished" two standard deviations below the mean for children considered well-nourished, by which criterion 25% of African and 70% of south Asian children are so affected. Poor children stunted at the time of weaning will probably remain stunted throughout their lives, with probability of reduced physical and oftentimes intellectual capacity. Protein-energy malnutrition raises susceptibility to infections, and is often found among children prematurely weaned and in families too poor to afford nutritionally balanced weaning foods (Scrimshaw 1993).

In all countries and communities surveyed, there exists compelling evidence that poverty and consequent inadequate access is the primary cause of undernutrition, many patterns of malnutrition and food insecurity. While science and technology have contributed and must continue to contribute to food supply and security, science alone cannot remedy all the world's nutritional ills, nor is there a simple single political panacea applicable and practicable for all situations. Food security, however defined, is a complex creature diverse and variable according to place and time. While remedies for certain specific deficiencies are understood and, with modest resources, realisable, extensive malnutrition of extreme chronic proportions calls for systematic analysis, the long-term commitment of capable people, substantial investments, and a degree of political will not yet evident among governments, donor and development agencies.

INADEQUACY AND IMBALANCE IN THE USA

In the international literature, most attention is given to hunger and malnutrition among developing nations of Africa, Asia, the Middle East and Latin America. The United States together with other affluent OECD nations are generally believed to control resources more than sufficient to satisfy the needs of all their inhabitants. In overall gross terms such is the case. Nonetheless, among North Americans and Europeans many melancholy incidences of malnutrition, of chronic and transient hunger among poor and disadvantaged peoples are reported. Many such reports emanate from non-government and social agencies and, in light of the expense and resources involved, it is not surprising that very few are of a scale that provide a comprehensive account of the magnitude and precise causes, or to what extent malnutrition and extreme hunger may be chronic or spasmodic.

No nation has given greater attention to or published more on human nutrition than the United States. From the time of the depression in the 1930's when U.S. farmers produced commodities that many citizens were too poor to buy in the commercial market, the alleviation of hunger has occupied the attention of many Americans. Few, if any, governments have instituted more programs than the United

States. A brief review of U.S. legislation, as it relates to hunger and malnutrition, is relevant to issues of food security.

United States Government Legislated Programs

Since the early 1930's successive U.S. governments have enacted various policies and interventions to alleviate hunger among the poor. In 1930 the Department of Agriculture and the Federal Relief Administration distributed surplus agricultural commodities as food relief through the Federal Surplus Relief Corporation.

In 1933, Congress created the Agricultural Adjustment Administration to control farm prices and production, and the Federal Surplus Relief Corporation to distribute surplus farm products to needy families. From 1939 to 1943 the Federal Surplus Commodities Corporation, under USDA, carried out an experimental food stamp program. In 1946, the National School Lunch Program began. In 1955, USDA estimated that, on average, low-income families spent a third of their disposable income on food. In 1964, Congress established the national Food Stamp Program and the Social Security Administration set the poverty line at 3 times the cost of USDAs lowest-cost Economy Food Plan. Since 1969, the poverty line has been adjusted to reflect changes in the Consumer Price Index.

In 1966, the Child Nutrition Act was legislated and President Johnson formulated a Food for Freedom Program. In 1968, the Senate established the Select Committee on Nutrition and Human Needs to guide the efforts to alleviate hunger. In 1969/70, following several state and pre-school nutrition surveys, a report "Hunger USA" recorded various evidence of malnutrition among poverty-stricken children. In 1969, President Nixon convened a White House Conference on Food, Nutrition and Health; and USDA created a Food and Nutrition Service to administer federal food assistance programs. In 1972, Congress authorised a Special Supplemental Food Program for Women, Infants and Children, and a Nutrition Program for Older Americans under the Older Americans Act. In 1981, USDA started the Temporary Emergency Food Assistance Program (TEFAP). In 1984, a President's Task Force on Food Assistance found little evidence of widespread undernutrition but declared that for hunger to exist in the USA is intolerable. In 1988, the U.S. Surgeon-General's Report stated that no American should be deprived of access to a satisfactory diet. Congress passed the Hunger Prevention Act which expanded eligibility for and benefits from the Food Stamps, Child Care and TEFAP programs.

The food assistance programs provided by successive U.S. governments for Americans who, for one reason or another appear unable to provide nutritionally adequate diets for themselves have been subjected to adverse criticism by more authors than can be considered in this publication. Food is a component of social assistance to the poor, the homeless and the elderly in many OECD countries. Experience in the USA, though by no means exactly typical of all others, is indicative of the complexity of reliably identifying and satisfying those in need and deserving of supplementary food assistance.

Poppendieck (1992) contends that despite the emotional response to hunger in the USA, during the Great Depression and more recently, no satisfactory solution has been formulated or implemented. The New Deal did not provide adequate defence against hunger in a land of plenty, a time when there was want in the midst of plenty: "want begot by plenty". The Roosevelt administration's food assistance program, based on the procurement and distribution of surplus agricultural commodities laid the political basis of food assistance up to the present time.

Poppendieck discusses measures to control commodity surpluses by coincident policies of reducing land under cultivation, ploughing under of crops and slaughtering of livestock that could not be marketed at profitable prices. In 1933 the Agricultural Adjustment Administration forestalled an impending glut of pork by slaughtering several million young pigs and pregnant sows. Those too small to be smoke-cured or canned were converted to liquid tankage used for fertilizer. Edible products were distributed to the unemployed by the Federal Emergency Relief Administration (FERA). Soon thereafter, an independent agency, the Federal Surplus Relief Corporation (FRSC) was chartered to administer an expanded program in which surplus commodities were distributed to the deserving poor.

Studies by the U.S. Society for Nutrition Education

During a National Conference on Social Work held in Chicago in 1934 the then Undersecretary of Agriculture, Rexford Tugwell is reported to have said: "FRSC helped the Department of Agriculture to solve the paradox of want in the midst of plenty". Poppendieck is critical of the FRSC program on the grounds that the prime objective was to protect prices by preventing excessive surplus, and that even when most responsive to the needs of poor people, FRSC purchases were small in relation to need. ". . . commodities represented an average increment of a little over 10% of the relief food allotment or less than one-fifth of the gap between the food expenditures of relief families and those of low wage workers". Poppendieck considers that the program did more to ease the political conscience than to satisfy the needs of the poor.

Poppendieck suggests that contemporary programs of food assistance such as the government's TEFAP and voluntarily operated soup kitchens are not an adequate remedy for the inequalities and deprivation caused by contemporary economic forces and widespread unemployment. Food aid and assistance entirely dependent on the distribution of commodities surplus to market demand will not satisfy the long-term nutritional needs of the indigent and dispossessed.

The U.S. Society for Nutrition Education advocates that the hungry be provided with adequate food and taught how to manage food resources to ensure their good health. Testifying before the Senate Committee on Nutrition and Human Needs in 1972, the SNE said: "every child. . . by the time he/she is 18 should have sufficient knowledge (about nutrition) to be able to make adequate food choices" (Chapman 1992).

Clancy and Bowering (1992) describe how the protracted need for the Emergency Food System (EFS), the expansion of soup kitchens, food banks and other facilities to serve the persistent needs of indigent people, have encountered difficulties in finding the necessary volunteers and food resources. Because of sparse insufficient quantitative data on the nutritional status of low-income households, and the extent of their dependence on emergency food, community nutritionists have difficulty agreeing upon the true nature and magnitude of poverty-induced hunger. Food security for all Americans cannot be realised until there is scientific data adequate to understand the incidence and scope of hunger and the underlying causes.

Clancy and Bowering report results of studies of people in New York State who, at least partially, depend on the emergency food system. They indicate that food insecurity has risen during the past decade; that in several households income, though above the poverty line, was too little to provide adequate food. The poverty line was arrived at following a USDA 1955 survey that determined average family expenditure on food to be approximately one-third of income. The estimated cost of the USDA Economy Food Plan is multiplied by a factor of three to derive the poverty line. The estimate is adjusted according to changes in the Consumer Price Index. Clancy and Bowering contend that since 1955, Americans spend proportionally less on food and more on shelter, hence the poverty line estimate is too low. Clancy and Bowerie (1992) and O'Hare et al. (1990) consider the present poverty line as inaccurate but report, among several who have studied the matter "no consensus on what a new definition should or could measure". Among the Clancy and Bowering sample the median income was 77% of the poverty level, median expenditures for housing and food were, respectively, 52% and 24% of income. Many depended on food stamps and commodity assistance programs.

Nestle and Guttmacher (1992) state that, during the 1980's, as a result of economic and social changes, large numbers of Americans, unable to afford an adequate diet, have become dependent on food assistance. Over the same period some responsibility for food and income assistance has shifted from federal to state and private organisations, neither being endowed with resources commensurate to the demand. Legislation enacted in the early 1980's restricted benefits designed to cushion vulnerable people from economic hardship. Emergency food services reported rising demands from "new poor" children, unemployed youths, low-income families and mentally ill people released from protective institutions. A report from the Food Research and Action Center (FRAC) in Washington DC, lists some 250 reports on hunger in North America since 1970 and reflects on the unfortunate consequences of reduced commitments to hunger alleviation by the Reagan administration (FRAC 1991).

The report of a Physician Task Force on Hunger in America (1984) discussed hunger in economic terms, stating that people with incomes below the poverty line, with inadequate food assistance, were at risk of hunger, by which criteria some 12 million children and 8 million adults were so affected. Matsumoto (1989) estimated that federal expenditures on food assistance exceeded $21 billion annually and benefited more than 40 million people.

Nestle and Guttmacher (1992) conclude that a paucity of data on the prevalence, intensity and pervasiveness impede the formulation of a coherent federal policy for hunger relief; that hunger is chronic and results from poverty not shortage of food; at greatest risk are those whose income is too low to provide such basic needs as food, shelter and clothing; the federal poverty line is an unreliable index of hunger not taking account of all basic needs and regional price disparities.

The social welfare system does not provide an adequate safety net; private charities that offer food assistance cannot satisfy the total national need. Freedom from hunger requires relief from poverty, which in turn calls for employment rewarded by an adequate minimum wage. Hunger must be tackled with due regard for other social inequities: underemployment, unsatisfactory housing and education.

Parker (1992) explains that while many elderly Americans may be at nutritional risk, there is insufficient biochemical, physiological and clinical data to indicate what, if any, specific changes in dietary patterns should be recommended as people become elderly. Family meals are social occasions. The disposition of elderly men and women who live alone to neglect themselves is not unusual. Parker reports results of a study in which some 3600 men and women over 60 years of age living in 21 states were surveyed. The average age was 73, 67% were women and 54% lived alone; 20.2% had involuntarily lost weight during the previous month; 5.4% were without food for more than three successive days in the month; 28.3% had no one to provide help if they were sick and confined; responses indicated that 21.7% were at risk of chronic malnutrition. Only 36.5% of those who had too little money for food participated in a Food Stamp Program.

Cohen et al. (1992) described hunger among homeless people who rely on soup kitchens and communal shelters. Though the meals provided, if eaten regularly, would furnish sufficient calories they tended to be low in fruits, vegetables and dairy products. From a survey of facilities that collectively provide 321,000 meals each day it was estimated that on average homeless persons received 1.4 meals per day. Cohen et al. (1992) considered that private charitable organisations provide a commendable and necessary service; they need more food and money to satisfy the many homeless who depend on them. Meals delivered by mobile food wagons would fill a particularly evident need.

Morris et al. (1992) examined food costs and availability among rural Americans to determine: the relative accessibility of rural and urban retail supermarkets; access to particular foods; costs of a USDA Thrifty Food Plan (TFP) market basket relative to food stamp allowances. Supermarkets and stores were surveyed in 33 nonmetropolitan counties. A TFP marketbasket of 77 low-cost foods in 31 food categories was priced in each store. USDAs estimated cost of a market basket was

$75; the average costs in supermarkets and small/ medium stores surveyed were, respectively, $81 and $102, from which data an increase in food stamp allotments and a revision of USDAs market basket price is indicated.

Smith and Hoerr (1992) interviewed low-income single mothers to find out how many received food assistance. The average income of the sample surveyed was 81% of the federal poverty level for a family of three. Those most dependent on food assistance had larger families with older children. The results indicate the dependence of many single mothers with families on food assistance programs but the study raised more questions than it answered.

It is apparent that many Americans because of poverty, disability, other impairment or disadvantage are unable to provide for themselves adequate diets. The dimensions of hunger and chronic malnutrition seem difficult to diagnose even with America's unparalleled resources. To examine, comprehensively all of the complex issues that impinge upon nutritional deprivation, its extent, complexity and causes; to prescribe and enact effective means of eradication seems to call for investments greater than federal, state governments or private agencies are willing or able to commit. The U.S. Society for Nutrition Education and their partners deserve commendation for their thoughtful commentaries and encouragement to continue their studies of hunger among America's deprived peoples.

The extensive and well-documented history of programs intended to diagnose, quantify and alleviate chronic hunger and malnutrition in the USA illustrates how even the most prosperous and well-endowed of the world's nations finds difficulty in assuring food security for all of its citizens, especially those least able to fend for themselves. Though much of the American experience is particular to that nation's political and economic circumstance and disposition, the record of the various U.S. programs and their consequences merit critical examination by agencies and jurisdictions responsible for diagnosing and designing food security policies and programs for other communities. The present weariness with well-doing and disenchantment with aid to the poor now apparent in Washington may well be influencing other affluent nations and conspiring to the general decline in support for international development in general and investment in agricultural development in particular.

CHAPTER 4. SUSTAINABLE AGRICULTURE

Recent years have witnessed an immense outpouring of published literature under the general heading of "Sustainable agriculture". In preparing an invited report on this subject for the UN Conference on Environment and Development in Rio, a Scientific Committee of the International Council of Scientific Unions (ICSU) reviewed some 300 publications related to sustainable agriculture (CASAFA 1991). Among the many treatises the subject is addressed from various scientific, pseudo-scientific, biological, ecological, ideological and philosophical perspectives. The following are a sample of the publications referenced in CASAFA 1991: Agriculture Canada (1989); Altieri (1983); Boeringer (1980); Brown (1987); Bunting (1987); CGIAR/TAC (1989); Edwards et al. (1990); FAO (1984); Jain (1983); NRC/NAS (1989); NABC (1989); Rodale (1983); Reganold and Papendick (1990); Swaminathan and Sinha (1986); WCED (1987).

In broad principle a sustainable agricultural system is one that so manages the essential and contributing resources as to satisfy the needs of all who are presently dependent without compromising the needs of future generations. It is in trying to prescribe production systems of broad relevance and ecological adaptability that complexity and contradictions arise.

Geological and anthropological evidence available indicate that from the time of creation, the earth, its environment and creative inhabitants have continually changed, sometimes gradually and evolutionary, sometimes in an extreme and disruptive fashion. Some observers believe environmental change is accelerating, in large part because of human activities. Some undesirable changes are attributed to agricultural developments during this century.

Much of the literature on sustainable agriculture originates from the industrialised, relatively prosperous countries and relates to their systems of intensive production. In these economies, there is less urgency to produce more agricultural food and fibre than to produce with less disruption to the environment and natural resources. In contrast, poorer nations must produce more abundantly in order to feed their fast growing populations.

Though "Sustainable Agriculture" seems the most favoured descriptor, there are others intended as synonymous or with similar connotations. The following, gleaned from the literature display many disparities in prescribed policies, purposes, priorities and practices.

Alternative: Strictly speaking, this describes any production system different from what is commonly practised. In some instances it implies ecologically benign systems less dependent upon agro-chemicals and other manufactured inputs, more reliant on nutrient cycling, biological nitrogen fixation, and the natural enemies of pests and pathogens; production systems that conserve and protect soil and water quality and biological diversity; profitable systems that respond to natural cycles and that integrate beneficial biological interactions among crops, their pests and pest predators.

Biodynamic: Systems that make significant use of compost and humus to benefit soil structure and fertility.

Ecological, Eco-biological, Socio-ecological: Stated to be based upon agricultural ecology and human ecology; protective of the environment, discourages use of chemical fertilizers, herbicides and pesticides.

Low input: Least possible use of agro-chemicals, fossil fuel energy, mechanisation, commercial certified seed and pedigree livestock. Greatest dependence upon on-farm breeding and seed multiplication, animal and green manure and compost.

Low input sustainable (LISA): Reduced production costs and increased profit by decreased dependence on off-farm resources. Productivity is increased by more efficient use of certified seed, energy, mechanisation and other purchased inputs; by soil and water conservation, conservative tillage, multiple cropping mixed with livestock husbandry, systematic recycling of manure and crop residues.

Organic: Based upon recycling of nutrients from on-farm resources. Proscribes agro-chemicals. Described in one text as "A precursor of sustainable agriculture".

Regenerative; Permaculture: Similar to Biodynamic and Organic.

These few brief extracts from a copious literature simply illustrate the many contradictions and discrepancies among concepts and definitions of sustainable agriculture. Each may be applicable in an appropriate set of conditions. No one can satisfy all circumstances.

Though some common themes appear, the definitions and descriptions of "Sustainable Agriculture" are at best general. The literature conjures up Humpty Dumpty's assertion: "When I use a word, it means what I choose it to mean, neither more nor less" or the clergyman who said he could not define sin but he sensed it when it was near. Several refer to satisfying human needs but, apart from an occasional reference to post-harvest losses, there is almost exclusive concentration upon production systems within the farm perimeter, with little concern for what happens between farms and consumers.

"Sustainable" has many interpretations and cannot be globally defined in terms of specific technologies or practices. The operating variables among agro-ecologies are almost infinite. The following purposes of "Sustainable Agriculture" are extracted from various publications:

1. Its objectives are to generate agricultural products adequate and acceptable in quantity, variety and quality; maintain environments favourable to humans and other organisms; prevent pollution of surface and ground water; protect wildlife and animal rights; prevent despoiling and degradation of fertile land by erosion, urban spread and other non- agricultural exploitation; government control of land and ownership to prevent misuse; maintenance of infra-structures essential to production and marketing of agricultural goods.
2. It seeks an orderly, harmonious balance among resources available, protection of the environment by systems that satisfy dependent human needs, social and economic interests.
3. Copies natural eco-systems but increases yields and productivity. Favours long-term conservation over short-term excessive exploitation. Seeks to improve photosynthesis and nutrient recycling. Ensures a rational compromise between high yield and long-term biological and ecological stability. Protects and takes maximum advantage of genetic diversity by collection, classification and conservation of macro- and micro-organisms beneficial to farming and surrounding environments.
4. Ecologically sound, economically viable, socially just and humane.
5. Continuous profitable farm production which preserves the natural resource base. Sustainable is not synonymous with low-input or organic since the available organics may be inadequate to maintain fertility particularly in poor soils. Carefully regulated pesticide application can be more efficient and cost effective than cultural or biological control of pests.
6. Farming systems that are profitable and environmentally safe, that produce adequate amounts of high quality food and protect the natural resource base.
7. Productive systems that raise crops and livestock in higher density and more rapid succession than natural ecosystems which can only sustain hunters and gatherers not farmer/cultivators. Modern farming systems take out more soil nutrients than natural ecosystems, nutrients that must be restored if production is to be sustainable.
8. Complex systems to reverse present serious trends in European and North American agriculture: high energy costs, depletion of fossil fuel resources, contamination of ground and surface water, soil erosion, declining returns to farm investment, excessive subsidies, concern for risks to human health and wild life habitats. In 1980 the U.S. Department of Agriculture estimated that about 1% of all American farmers were practising non-conventional "sustainable" agriculture.
9. Combines traditional conservation with modern technologies. Inputs purchased from outside are replaced by natural on-farm or near-farm resources: solar and wind energy, biological nitrogen fixation, cultural and biological pest control. Mono-cropping is replaced by a planned succession of a variety of crops.

10. Seeks to realise several main objectives: Efficient but often complex diversified production systems; conservation of ecology and natural resources; protection of surrounding environment, of ground and surface water, and of flora and fauna; conservation of non-renewable energy resources; harvested crops free of pesticide residues; adequate and dependable farm income.
11. Multiple in place of mono-cropping; mixed crops with livestock in patterns and densities compatible with agro-ecological conservation; cultivation of food, forage and cover crops, with live mulch and green manure; recycling of nutrients from manures and residues; cultivation to minimise soil erosion; more systematically controlled use of agro-chemical fertilisers and pesticides.

Being derived from the Latin "*sustenire*", "to sustain" literally means "to uphold or to keep from falling". This may suggest maintenance of the status quo: that sustainable agriculture is a static concept. History illustrates that with little or no population increase various traditional systems of agriculture were sustained over many centuries. If the fallow period between successive cultivations is long enough for vegetative ground cover to regenerate, shifting cultivation systems can sustain limited dependent communities but not the original mix of natural organisms. "*Sustenire*" is an active transitive verb: someone must act "to uphold and keep from falling". The actions needed will change as the needs and resources of agro-ecologies change.

Many observers question whether proscription of agro-chemicals is practicable and suggest that fertilizers and pesticides best be used to complement other practices. Farmers on the U.S. Great Plains whose topsoil blew away during the 1930's used little if any chemical fertilizers, pesticides or herbicides. The same is true of lands in the semi-arid tropics turned into desert by overgrazing and overcultivation: by repeated planting and harvesting with little restoration of nutrients.

Most sustainable systems are specific to locations, to prevailing agro-ecological and socio-economic conditions; they indicate caution against foreign agencies seeking to impose upon resource-poor nations what appears desirable in North America and Europe. The urgent need in many developing countries quickly to increase food and fuel production is often in conflict with an ideal state of ecological conservation.

An experienced Indian scientist wrote recently: "You hear a call these days from parts of the First World which urges a return to low-input technology. . . . that is the worst advice that can be given to farmers in Third World countries. With the population pressures they have, they can't afford to go back to low input technology. That is what they had for thousands of years. Farmers in developing countries need more sophisticated technologies than the ones they are using today" (Jain 1983).

A definition quoted in several publications reads: "Management and conservation of agricultural resources to satisfy changing human needs while maintaining or enhancing the quality of the environment."

Harwood (1990) has a written a thoughtful historical review of what are perceived as sustainable production systems and their underlying philosophies. He writes: "The word 'sustainable' implies steady state. If one seeks a steady state situation, one must look over horizons to some distant goal." Harwood goes on to say that the literature reveals as many ideas as there are authors, and that a wide diversity of understanding, a lack of hard data and consensus on resources, environments, future technologies, and the interrelations among all related factors make prediction of what will be widely sustainable, in any precise sense, a futile exercise. Harwood suggests as "a workable definition: agriculture that can evolve indefinitely toward greater human utility, greater efficiency of resource use, and a balance with the environment that is favourable both to humans and most other species". Here again a concept with which, in overall principle, few would disagree, yet open to many and diverse practical interpretations.

Harwood reviews the history of biodynamic agriculture, humus farming- that inspired many disciples and publications from 1930 onwards: organic, biological, ecological and regenerative farming systems, each in various ways seeking soil fertility through recycling of organic waste and residues rather than by recourse to synthetic chemicals. Despite enthusiastic advocacy of these conservative intentions, rural to urban migration, availability of cheap fertilizers, pesticides, machines and fuels, propagation of high-yielding hybrid crops encouraged American farmers to adopt intensive production technologies using industrial inputs rather than systems purported to be more ecologically benign and "sustainable".

An extensive analytical review of "alternative systems" for U.S. farm production was assembled by a committee of the Board of Agriculture of the U.S. National Research Council (NRC/NAS 1989). The study was prompted in part by concerns in the U.S. Environmental Protection Agency which identified agriculture as a major source of surface water pollution. Other concerns relate to soil erosion, evolved resistance to pesticides in food crop pests, pesticide residues in foods, and the rising costs experienced by U.S. farmers, many saddled with a heavy load of debt. The NRC report states that U.S. farmers are examining alternative practices with a view to reducing production costs, conserving natural resources, and protecting the health of consumers.

"Alternative agriculture" is a broad term that embraces production systems and technologies different from those considered conventional. The alternatives discussed by the NRC committee seek to encourage such natural processes as nutrient cycling, biological nitrogen fixation, beneficial pest-predator relations; along with reduced dependence on off-farm inputs, more effective use of biological and genetic potential in crops and livestock, protection of soil and water by design of cropping patterns that take account of physical limitations, under management practices that conserve energy, biological and all other essential resources. The NRC committee conceives "alternative" not as a single system but a spectrum that can range from "organic", using no processed chemicals to practices that use agro-chemicals prudently. Several of the alternative practices considered by the NRC committee are discussed later in this paper.

Development agencies delight in "buzz words" which they often fail to define with any degree of precision. What is sustainable depends upon what or who are to be sustained, at what level of sustenance, for how long and with what resources. A sustainable system is specific to a particular location and will change as conditions change.

Agricultural Resource Management

In this context, "Management" covers many levels of authority and activity from policies defined by international agencies and national governments to decisions made and technologies adopted by farmers. Changing human needs extend from the near into the indefinite future, the needs becoming less predictable as the horizon recedes. Needs will comprehend those of particular nations and communities for adequate food, fuel, fibre and other products; farmers needs for a just and equitable return for their investment and labour; industries' needs for raw materials suitable in quantity and quality to be processed to meet market demands; the need to provide gainful farm and agro-industrial employment. Quality of the environment is difficult to define except in broad generalities. It implies being careful of natural ecosystems and conserving and restoring resources sufficiently to satisfy the needs of present and future generations (CASAFA 1991).

There are inconsistencies among interpretations of "resources". Some writers restrict sustainable agricultural resources to those within farm boundaries. Others take in all biological, chemical, physical, naturally occurring, manufactured, economic and human resources that enter into agricultural production, processing and distribution, illustrating the need for a broad comprehensive approach to resource allocation and management.

Various authors raise critical questions in their appraisals of proposed sustainable systems: What are the specific purposes of the system recommended? Who or what is to be sustained? At what standard of living are particular populations to be sustained? What patterns of change and rates of growth are assumed? What environmental or climatic changes are anticipated? What are the assessed costs and risks and by whom will these be borne?

These questions are by no means satisfactorily addressed in many of the texts reviewed, with more attention given to technologies than to social and economic issues. Opportunities, constraints and resources vary greatly among locations and communities. Resource management must therefore be considered holistically, and as it relates to each specific entity.

The immense disparities in conceived priorities between the wealthy and the impoverished nations make clear that there can be no globally applicable policies or courses of action. Questions of cost and risk will bear much more heavily upon poor Africans, Asians and Latin Americans than upon those European farmers who receive substantial subsidies. Any increased prices in the market place, consequent upon significant changes in production systems, will probably be more tolerable to West Europeans and Americans than to the developing nations' consumers.

Under the most favourable of conditions, farming incurs immense risk. However defined, agricultural systems must be sustainable or they will collapse. Over many centuries agricultural technologies have evolved to satisfy changes in need, demand and opportunity. Future proposed change must take calculated account of risk, the nature and magnitude of risk and by whom the risk is to be borne. A risk tolerable in one place could be disastrous in another. It is to be hoped that systems which appear to be ecologically benign and, in the long term, sustainable will first be studied systematically and prudently before being urged upon farmers. This demands far greater investment in research than is presently provided. It requires sustained support for national and international agricultural and food research agencies and services, and for the facilities needed to ensure safe preservation, processing and distribution between producers and consumers. Serious discussions of sustainable agriculture and food security must embody consideration of all these essential integrals.

AGRICULTURAL POLICIES

A report to the U.N. Conference on Environment and Development emphasised that relative to food security and sustainable agricultural development, scientific research and technological innovation must be consistent with food and agricultural policy (CASAFA 1991).

Agriculture is strongly influenced by government policies, policies that determine prices, terms of trade, incentives as subsidies, credit and insurance protection. Among poorer countries, budgetary constraints discourage long-term planning, and agriculture is often assigned a relatively low priority. Too little political and financial support forces farmers to adopt production practices that offer quick returns, practices that extract more than they restore to the soil. During the 1980's, in Africa prices paid to farmers fell by 17%. Over the same period prices to Asian farmers rose by 11% and in Latin America by over 15%. Price and other incentives tend to be more generous for exportable commodities than for local foodcrops. Smallholders, women in particular, customarily have restricted access to credit and extension services.

Few governments enact ecologically protective legislation: to control land use; to protect forests from wasteful exploitation, lakes, rivers and aquifers from pollution and overdrafting; to encourage systematic land improvement and irrigation management (CAEFMS 1990). Control of the water resource is often spread across several uncoordinated ministries so that misuse is difficult to monitor. Few governments seem concerned with agricultural production for future food security; with protection of essential natural resources in their own and neighbouring lands. For nations too poor in financial, institutional and human resources to implement desirable policies, international agencies could helpfully provide policy advice and monitoring services on a regional basis.

Ecologically sound policies cannot rely simply upon technical issues but must take account of many social, economic, legal and institutional considerations. Nor should they be so strongly environmentally biased to put at risk essential food supply and security. Policies may embody various options: regulatory control; incentives by subsidies; taxation disincentives; property rights regulations which define liability and compensation for damage. Several economists recommend, where feasible, well defined property rights to encourage resource conservation, the essentials being: identification of ownership and specific assets; the owners having legal protection against misuse by non-owners; the assets owned being necessarily legally divestible and transferable. This may not be simple. If, for example, emissions from a factory cause damage to particular farm crops, the applicability of a property rights action is relatively straightforward. Where indeterminable amounts of soil and contaminated run-off water from several farms enter ground or surface waters, the process by which property rights compensation may be determined and imposed is less apparent (CAEFMS 1990; Hulse 1991a).

To bring about desirable changes may require the dismantling of long established policies that subsidise intensive rather than conservative production systems. Such policy reformulation will likely meet with opposition from those who have benefited most. New policies will need to be consistent with the interests of farmers, consumers and all other concerned and affected people, and implemented with enough political strength of will to discourage evasion by vested interests.

Policies Unfavourable to Resource Conservation

A published commentary on U.S. agricultural policy states: "Between 80 and 95% of all acreage producing corn, feed grains, wheat, cotton and rice are in federal commodity programs. All acres enrolled in the federal commodity income and price support programs are subject to crop program rules. Any practice that reduces the acreage planted will reduce the acreage eligible for federal subsidies for the next five years. If a farmer rotates the base acreage for one year to a legume, program payments will be reduced by 20% for 5 years" (NRC/NAS 1989). Other restrictions may discourage diversification from a single crop to intercropping or crop rotations.

In the USA, under the 1973 Agriculture and Consumers Protection Act, farmers are guaranteed a per bushel price. Payments are tied to size of crop base and the bushels harvested. A large crop base and high crop yield qualify for highest payment (NRC/NAS 1989).

U.S. policies have committed large areas to intensive mono-crops, often to the detriment of soil fertility and stability. Subsidies to European farmers encourage production surpluses which when dumped onto world markets depress prices and discourage food crop production in poorer developing countries.

Grading standards for some horticultural products prohibit such extraneous materials as insect fragments, leading growers to apply more pesticide than is necessary to control serious plant physiological or nutritional impairment. Excessive spraying increases production costs, pesticide residues and the rate at which target pests evolve immunity. In the USA stricter standards are applied to new pesticides than to pest control methods approved before 1971. Little evidence is published of governments having systematically determined economic returns and benefits from alternative pest control practices. Nor do governments anywhere seem disposed adequately to finance research and training to develop and demonstrate reliable systems of biological control and integrated pest management that are less dependent upon pesticides.

Though sustainable agricultural production, protection, processing and distribution should always be compatible with environmental protection and conservation, the means and policies to realise these ends will differ widely among countries, communities and ecologies. Neither means nor policies should be arrived at in haphazard fashion. All governments need access to professional policy research competence to assess and balance needs against risks, resources and constraints; to rationalise immediate and predictable future demand for food security with sound ecological stability.

Policies that encourage shifts from intensive to diversified systems of production may result in higher prices for some staple commodities and processed foods. It would seem politically expedient to determine in advance what price increase consumers can tolerate in the interests of ecological conservation.

It is neither wise nor just for wealthy nations, either by political persuasion or through aid channels, to try to impose what they conceive as desirable agricultural and environmental policies upon developing countries. Though certain guiding principles and broad objectives may be generally agreed upon, the needs, resources and constraints will be particular to every nation and community. Each must, therefore, decide and apply what best satisfies their conditions, needs and aspirations.

Publications on sustainable agriculture and agricultural policy are largely written from a North American or European perspective, many by observers whose personal welfare is not at risk. Policies and practices deemed desirable and affordable by affluent nations can do damage when imposed upon poorer nations, where money, land and other resources are severely limited, and where populations are rapidly expanding.

Agricultural Policies in the USA

American agricultural policies and their influence on farmers are discussed in many publications from the U.S. Department of Agriculture, by a National Research Council Committee (NRC/NAS 1989) and by Busch and Lacy (1984). The NRC Committee comments on the visible influence of U.S. commodity programs on the decisions made by farmers, and states that "In 1986, total federal farm program outlays equalled nearly 50% of net cash farm income. . . (which) reached record levels of. . . $46.3 billion in 1987 due in part to the large subsidies paid to most growers who participated in federal commodity programs. "The NRC Committee goes on to illustrate how government programs and policies largely decide farm management practices and what crops will be grown, since prices under commodity programs are often far above international market prices. The manner in which farmers use their land is strongly influenced by government program regulations and incentives.

Federal commodity programs exist ostensibly to stabilise, support and protect crop prices by direct payments to farmers, subsidising exports and either encouraging larger plantings of particular commodities or persuading farmers to take land out of production. The U.S. Department of Agriculture appears to have recognised that policies are developed as a patchwork of individual programs rather than in the form of overall coherent strategies. USDA published the following observation: "Farm policy has always tended to follow events and changes rather than to anticipate and lead them. . . . There is little doubt that some of the programs that have resulted from this ad hoc, crisis oriented policy making have subsequently exacerbated problems of farmers, or produced unintended and unwanted consequences for the farm sector as a whole." The report states that design of farm policies based upon systematic studies rather than by adjustment of existing statutes has been rare.

The NRC report discusses how tax policies can influence agriculture: favourable capital gains legislation offers incentives to purchase wet lands and rangelands at low prices, to convert them to erodible crop land for later profitable sale with limited taxation on the capital gain realised. The NRC Committee, Busch and Lacy (1984) among others point to agricultural policies that are inconsistent with desirable objectives of sustainable agriculture and resource conservation.

LeVeen (1984) and Freudenberger (1984) describe how international events influenced United States' policy during the 1960's, 70's and 80's. A period of stimulated high production during the 1960's was followed by the so-called global energy-and-food crisis of the 1970's which led to the incentives that encouraged over-production in the 1980's, with farmers incurring large debts to finance the larger harvests demanded of them. The despondent forecasts that emanated from the World Food Congress in 1973 proved unduly pessimistic and caused farmers to produce food commodities in quantities that depressed prices in world markets. The agricultural depression of the early 1980's was to older farmers reminiscent of the U.S. agricultural economy in the 1930's. Then came severe droughts in many grain growing ecologies causing a heavy draw-down on grain stocks and an escalation in prices.

Fluctuations such as these apparently do little to remind those who decide agricultural policies that agriculture is inevitably subject to unpredictable climatic variability. As pointed out by LeVeen (1984) the extreme perturbations of the 60's, 70's and 80's reveal that even the USA, the most advanced of nations, has a limited capability to control or to stabilise production or world prices. Relatively modest shortages or surpluses can drive prices up or down. Sudden changes in production systems and output, in response to demands by environmentalists could raise food prices to levels that would restrict access to poor people.

Policies that Bear on Developing Nations' Food Security

In a far-reaching review of future prospects for agriculture around the world, with particular concern for Third World countries, an FAO report makes the following comment: "On the issue of the role of the State in economic activity in general and in agriculture in particular the debate on policy reforms which emphasise efficiency and private initiative had - at its inception - more to say about what governments should not do than what exactly they should do. . . . With the passage of time, policy approaches are modified and fine-tuned as more lessons are learned as to which policies work and which do not, and why." (FAO 1993).

In countries poor in resources and infrastructures, price incentives when applied to traditional technologies are often insufficient to stimulate agricultural growth. Incentives that rely not only on price mechanisms, such as investments in relevant research, the provision of necessary production inputs, establishment and maintenance of rural infrastructures, transportation and marketing systems, will often prove more encouraging to farmers than sole reliance on price control.

Developing country policies are strongly influenced, not always for good, by donor agencies who offer technical and economic assistance, by international and regional banks who furnish loans. In the past most donor emphasis has been on raising agricultural production with little policy attention or investment given to marketing, preservation, other services and amenities essential to safe, economic delivery from farm to consumer. Delgado and Mellor (1987) comment that long-range effects of price incentives differ among countries and regions, with partial dependence on the availability and quality of land, water and other essential resources. Responses to price incentives need careful analysis and assessment in relation to other policies and factors that impinge on agricultural production and marketing.

A conclusion drawn in the FAO report is that piecemeal policies are often counterproductive; it emphasises the complementary relation of agricultural policies to overall macroeconomic, fiscal and monetary policies. The FAO report discusses how national policies are influenced by and in turn affect international commodity trade. For the most part international commodity agreements have not effectively prevented wide price fluctuations particularly detrimental to poorer nations. Barriers to trade in commodities and processed agricultural products, imposed by affluent industrialised nations, place severe constraints on Third World countries whose main source of income comes from agriculture (FAO 1993).

Koppel (1984) has written a thoughtful commentary on policies in relation to international trade, and to food security in the USA and elsewhere. At the end of the last century, 1/3 of American wheat and 1/3 of the cotton crop was exported. In recent years close to 60% of the wheat, 40% of soybeans and 30% of maize harvested are exported. During the 1970's at least 85% of increased production from American farms was to satisfy a 3-fold rise in exports. Dilation in international trade has led to volatility in demand, supply and prices with unfortunate consequences for farmers in many lands and the natural resources upon which they rely.

The political response to perennial excess production capacity has been to enforce reduction of land under cultivation. The U.S. Food Stamp Program, intended to alleviate distress among poorer Americans, has served more to lower storage costs than to raise or diversify internal consumption. Whether higher exports of U.S. surplus commodities truly respond to an identified market demand, or whether surplus commodities are pressed upon poorer recipients is a moot question. While only a few countries are major exporters of food commodities, over 50 are frequent importers of food grains.

Imports into regions with unreliable rainfall vary from year to year with a responsive instability in international markets and prices. To ensure food security for people whose home harvest is unpredictably variable, reserve stocks are essential. The cost of maintaining reserve grain stocks is considerable and excessive stocks may destabilise international prices. To provide reserve stocks that are economically sustainable calls for more reliable information and data, more creative critical logistical and economic analysis than appears to have been brought to bear in the past.

By and large, export policies are integrals of overall foreign policy objectives. Agricultural policy is often more influenced by fiscal and monetary concerns than by considerations specific to agriculture. During the 1950's, poorer LDCs and members of the Soviet Bloc accounted for 18% of international agricultural trade. By 1983 the proportion had risen to 43%, most of the increase being attributable to higher demand from the Soviet group. World agricultural trade expanded in the 1970's faster than commodity production. Since the 1950's, grain production to feed livestock has risen from 20 to 40% of world total, as a result of which sales of high quality wheat for human food have been overtaken by coarse grains and oilseeds for livestock feed. During the 1960's, 90% of American grains were exported on concessional terms; the proportion fell to 16% by the late 1970's and to barely 3% by 1982.

In part because of a collapse in the International Wheat Agreement during the late 1960's the U.S. abandoned its large grain reserves which disrupted supplies and exacerbated price volatility. To what degree the rising demand for coarse grains for livestock feed, particularly among the prospering nations of SE Asia, will cause confrontation between U.S. domestic supplies and exports is uncertain.

Food Policy Formulation

Foods and the commodities from which foods are derived are perceived differently by different people. For subsistence farmers food crops are their immediate source of survival; for producers of cash crops they provide disposable income; politicians often regard food commodities as negotiable means to foreign exchange, as instruments of social welfare and political influence; to the military mind food can be a powerful strategic weapon. While it might appear axiomatic that food's primary purpose is to satisfy human nutritional needs, commodities conceived as instruments of international trade and political influence tend to over-ride considerations of the basic needs of poor people.

Tabucanon (1993) urges a rational balance between agricultural production and trade, and food security; trade and poverty alleviation being often in conflict. Tabucanon proposes an analytical framework described as a methodology of multiple-criteria decision making (MCDM). The MCDM framework considers (1) food production and allocation of resources; (2) marketing and distribution; (3) nutrition policies; (4) national and international trade; (5) food research and processing. These in turn relate to various other factors such as existing and alternative agricultural technologies; patterns of food consumption; land, labour, capital and other resources;market prices and pricing policies; essential infrastructures; budgets and administrative issues; variable agro-ecological conditions.

The hierarchical complexity of policy making is illustrated by water resource allocation which can involve various public, para-public and private agencies. In a particular Asian country, water supplied to farms from a reservoir constructed for hydro-electric power generation, passes through the jurisdictions of 11 government departments before reaching farmers fields.

Tabucanon (1993) sets out what might appear as self-evident truths: before problems can be defined and solutions decided, all alternatives must be understood and guiding criteria established. Formulation of food policies is complicated by the many criteria and objectives to be considered in order to decide which are most important. The degree of complexity ranges from a conviction that all listed objectives are considered indispensable with no trade-offs possible, a situation for which there is no easy solution;to one in which most are dispensable with many trade-offs possible. Since policy criteria and objectives are difficult to quantify and accurately define, food policies usually derive from a balanced assessment of the opinions of diverse interested people, many of which may be more in conflict than consonance. Food and agricultural policies interact with and are influenced by various other economic, technological, social and political issues; by ecological and demographic concerns. It is tempting to relate food policy to such isolated issues as population control, farming systems, soil, water and energy conservation. Economists may emphasise income distribution and price controls while agronomists give weight to production systems and technologies. It is particularly misleading to equate carrying capacity based on estimates of potential yield and production with consumption. These make little allowance for diversions, dislocations and losses between farms, markets and consumers.

Food policies may seek (1) self-sufficiency, (2) nutritional adequacy for all citizens, (3) expanding and diversifying agricultural economies to provide food, employment and income for rural and urban dwellers, (4) commodities for export, (5) agro-industries to preserve, process and distribute foods, adding value to agricultural commodities; others to provide materials and services essential to farmers.

Affluent nations with abundant resources can realise all of these objectives. Poorer nations may find difficulty in achieving any (Tabucanon 1993).

It is conventional wisdom among advocates of unregulated markets to disparage government interventions. Yet they clamour for government protection when foreign companies can deliver processed products at lower prices. Every nation, irrespective of economic status, has need of a rational food policy, food and agriculture being critical to the health, social and economic welfare of all people. Government policies provide for safe delivery to markets and consumers: by investment and taxations to maintain rural roads and transport systems; by regulations that control composition, quality and labeling of fresh and processed foods. Subsidies to farmers and price supports influence what is produced and what is purchased. Visible and hidden tariffs by impeding imports of foreign foods stimulate indigenous agricultural production. Standards of identity can be so specified as to restrict manufacture to particular indigenous processors.

Peterson (1993) summarises the findings of several earlier studies. The market price of agricultural commodities, relative to the cost of purchased inputs is substantially higher in affluent than in poorer countries. Among Third World countries, the ratio of output value to the cost of labour is roughly 3 times higher in low income than in high income nations. Cost and return relations bear on international trade and on national agricultural and food priorities. Farmers in developing countries who receive low prices cannot afford to diversify their production systems, or run the risk of adopting new technologies. Governments avoid the political perils of raising food prices by not paying farmers fair prices for their labours.

In July 1995, the International Food Policy Research Institute in cooperation with the U.S. National Geographic Society convened an international conference on "A 2020 Vision for Food, Agriculture and the Environment". The following are only a few of the many subjects of relevance covered: Global food projections; Socio-political effects of new agricultural technologies; Implications of trade liberalisation and regional integration; Projections of malnutrition, food security and insecurity; Agricultural growth in relation to poverty alleviation; Science applied to the food needs of sub-Saharan Africa.

Issues relevant to providing for future food demand are addressed in several extensive reviews (IFPRI 1995a). The perceived challenges and recommended courses of action are presented in a draft publication to be modified and finalised later in the year (IFPRI 1995b).

The following are no more than a brief and far from comprehensive abstract of some important recommendations:

- Agriculture and sound food and agricultural policies are critical to sustainable economic growth and the assurance of food security
- In the lowest income countries, agriculture provides half of GDP, 3/4 of employment opportunities; over half of export earnings.
- Governments in all countries must formulate priorities, policies and strategies to ensure food security well into the next century.

Donor and development agencies must provide sustained support to:

- Strengthen the capacity of poor countries to realise food security by improving food and agricultural research and extension systems; utilise and manage natural resources conservatively; raise standards of living and provide employment opportunities for poor people.

In most countries a strong governmental commitment to agriculture and food security is essential. Each country must identify functions for which government agencies are responsible; incentives and policies necessary to encourage desirable private commercial investments.

The IFPRI 2020 Vision recommends that governments

- Develop clear and transparent policies and strategies
- Assist the private sector in predicting investment opportunities
- Maintain exchange rates, monetary and fiscal policies to stimulate broad-based economic growth
- Seek improved access to international markets through bilateral and multilateral trade negotiations and regional integration.

Governments are urged to enhance the health and productivity of low-income people by supporting education and primary health care; by providing facilities for training in productive skills; enacting land reform and property rights legislation; providing access to credit and savings institutions; by monitoring the state of food security among low-income people (IFPRI 1995b).

These recommendations are broadly consistent with others published and reported by the ICSU Committee (CASAFA 1991):

- In the interest of public health, welfare and food security, every nation needs a clearly conceived and defined food and agricultural policy.
- National agricultural policies must be integrated and consistent with overall macro-economic, science, technology and social policies.
- Policies must be formulated holistically not piece-meal and, while flexible, remain consistent and not subject to frequent fluctuation.
- Policies must be cogent and consistent with political priorities, economic, social conditions and resources of each particular nation.

The Politics of Agricultural Trade

Solkoff (1985) consummately addresses the politics of trade in agriculture and food, describing and discussing some of the less benign and humane features of modern agribusiness. He cites examples to show how agricultural commodity policies and trade can be used to serve political ends and to punish other governments and nations of whose political ideologies the government that controls the commodity may disapprove.

During the early 1980's, an appreciating U.S. dollar, mounting debt in many countries, combined with an overall global recession constrained all international trade in agricultural commodities. U.S. grain stocks rose and prices were depressed. Stimulated by direct and indirect subsidies, the European Community's share of global agricultural commodity trade rose by 155%.

As stated above, international commodity trade is strongly influenced by international capital markets, monetary and fiscal policies, and exchange rates. A high U.S. dollar discourages exports to countries with depressed currencies. The as yet untapped potential capacity of the former Soviet bloc to expand grain production and harvest surpluses for export could, if realised, cause future perturbations in global commodity markets.

Although most U.S. grain is exported by private companies, the U.S. government is an active participant using political influence to favour acceptance of U.S. produce. Aggressive promotion of U.S. wheat after WWII was largely responsible for changing Japan from a predominantly rice diet to a substantial consumption of bread. More recently political pressure has been imposed to persuade Japan to buy American rice. Though the grain trading companies are considered transnational corporations, with no particular national allegiance, they, and the U.S. cooperatives whose exports run to billions of dollars, are quick to benefit from the incentives and persuasions that the U.S. government can offer to potential importers. It is estimated that about 20% of U.S. grain exports are by such Japanese companies as Mitsubishi and Mitsui who own and operate grain elevators in the USA (Freudenberger 1984).

U.S. agriculture has always been closely wedded to international markets. With the inexorable decline in family-owned and operated farms and the expansion of large scale industrial production units, whose first concern will be the price of their stock on Wall Street, it is probable that farm commodities will be sold where the price is highest irrespective of the consequences for American consumers.

During the early 1980's, while farm prices dropped by 4%, consumer food prices rose by 4%, indicating that most effective price control is exercised by those who occupy the middle ground between farms and consumers. The expanding distance, spatial and philosophical, between producers and consumers may not auger well for consumer-oriented food security, particularly among the poorest people (LeVeen 1984).

Comment on Food Policy Formulation

From what has been referenced, and from many other policy proposals, inclusion of which would exceed the limits of this publication, it is clear that as John Bunyan stated: "Every vat must stand upon its own bottom". While certain general principles of policy formulation are broadly applicable, each nation must define policies specific to its particular needs and resources. External advisers and donor agencies should not seek to impose their specific policy conventions upon other nations, particularly those less fortunate than their own.

Among the poorer nations there are many who lack the human resources and experience to diagnose needs and define policies appropriate to those needs. Two international agencies, FAO and IFPRI, demonstrably can offer exceptional experience in food policy formulation. Would it not be feasible for the two organisations cooperatively to provide an international food policy advisory service to assist nations whose indigenous resources are inadequate to formulate policies and programs essential to their long-term food security?

RESOURCES FOR AGRICULTURE

All national economies depend upon the availability and quality of natural resources. Resources essential for agriculture can be classed as (i) Internal: Those that exist and can be renewed within a farm's confines; (ii) External: Resources that must be purchased, contracted for or otherwise imported from outside the farm. These may be sub-classed as (a) Natural: biological and physical; (b) Socio-economic; (c) Human. All agriculture depends upon the availability and quality of land and water; the production systems and technologies practised determine for how long productivity will be sustained and whether surrounding ecologies are conserved, improved or degraded (CASAFA 1991).

Internal Resources

Land is of primary importance, the classification, characteristics, biological and physical condition being determinants of the farming systems, crop cultivations and livestock management that are feasible. Scope and limitations are influenced by (i) land type, its profile and slope: whether flat, steeply inclined or undulating; naturally wet or dry; with or without good drainage; (ii) soil quality: structure, depth, density, chemical and organic composition, acidity, water-permeability and retention. Efficient land and soil management restrict erosion, the degradation of structure, fertility, and depletion of plant nutrients.

Water is essential to the survival and health of all farm crops and animals. Production systems depend upon the frequency, intensity and distribution of rainfall; on surface and ground water within and near to farms. Prudent water management protects against pollution; conserves surface, fossil and ground water; prevents overdrafting of rechargeable aquifers and contamination by agro-chemicals and organic effluents.

Atmosphere and climate influence seasonal temperature cycles, periods of insolation, air quality, favourable and unfavourable climatic change.

Flora include: (i) natural vegetation, some useful for food, forage or fuel, some being unwanted weeds; (ii) cultivated crops; (iii) soil and airborne microorganisms, some beneficial to plant reproduction and health, others being harmful pathogens.

Fauna comprise: (i) natural wild life: (a) helpful organisms, such as insect pollinators; predators and parasitoids that attack crop pests; earth worms and other creatures that improve soil properties; (b) destructive organisms: arthropods and rodents which directly injure crops or livestock, or perform as vectors for pathogens; (ii) Livestock:(a) wild animals; (b) animals husbanded for food, fibre and/or on-farm work.

The sustainable carrying capacity depends on relations and reactions among all of the above and in most places is arrived at empirically, by perception and experience. When the load on a farm's carrying capacity exceeds the internal resources, supplementary external resources will be necessary. External resources are often needed to support a transition from simple traditional to intensive production systems with novel technologies. Ideally, the natural resource base on every farm should be regularly and thoroughly monitored. In most cases this will be impractical, the cost being too great, the technicians needed being too few. Consequently, detailed evaluations are usually confined to selected on-farm case studies and experimental sites sufficiently similar to be representative of a larger farming community. Some governments support weather monitoring and forecasting, some maintain land-use inventories and offer soil analytical services to farmers.

Among affluent nations farmers can call upon technical advice from suppliers of agro-chemicals, animal feeds and machines. Less prosperous communities must rely upon government extension services many of which are poorly financed and lack the staff and resources necessary to provide the technical support needed.

External Resources

Land and soil

Off-farm services can be contracted to change land profiles, to build terraces and water catchments, to plant trees as wind-breaks, to provide machines and equipment of too large a capacity or too expensive to be bought and owned outright by farmers.

Water

Irrigation is used to supplement natural precipitation, either year round or during low rainfall seasons. Whether from on-farm or off-farm water sources, capital investment is needed for powered pumps, pipes and canals, and for equipment to distribute water across cultivated fields.

Energy

Resource-poor farmers depend upon human and animal power and often burn crop residues and manure as fuel. Animals need forage and feed that must be grown or purchased. Mechanical power from electricity or fossil fuels is needed to drive machines for soil cultivation, harvesting, post-harvest processing and transportation, for heating and cooking. Fossil fuels are essential to the production of agro-chemicals, prices being largely controlled by the world price of oil. As farming systems become more intensive, demand for external energy resources inevitably rises.

Flora and fauna

Production and use of commercial crop insecticides, fungicides, herbicides and other pesticides has grown rapidly in recent years causing concern for human and ecological health from excessive use, a concern that has stimulated research in biological control using natural or exotic enemies, predators and parasitoids to attack crop pests. For livestock producers there are diagnostics, prophylactics and therapeutics for detection, prevention and cure of diseases; synthetic hormones and other biologicals to control fertility and growth (ILRAD 1990, 1991).

Economic Resources

In addition to cash income from sales of farm produce and off-farm employment, farmers can be assisted by various direct and indirect subsidies, by low interest loans and credit, by crop and livestock insurance, by guaranteed sales through marketing boards, by bulk buying of inputs through cooperatives. Most farmers need economic support as insurance against uncontrollable misfortunes. Extravagant subsidies cause distortions of commodity prices and of market supply and demand. Subsidies that stimulate excessive production in one community can do harm to others. Total direct and indirect subsidies to agriculture in North America, Europe and Japan were estimated at close to $250 billion per year in 1990. Direct subsidies in western Europe approached $50B with additional benefits of close to $90B from prices fixed well above rational market values (CASAFA 1991).

Governments tend to overlook disparities between social costs and farm economics. On-farm costs take into account labour, seeds, machines, agro-chemicals and other inputs, and implicit opportunity costs such as farm family labour and equity investment. Social costs can be imposed beyond farms by water pollution from erosion, organic nitrification and agro-chemicals.

Low food prices for the benefit of urban dwellers may depress prices to farmers who then increase production by erosive land-use systems. Land is overcultivated or overgrazed, soil is degraded and nutrients depleted. Unable to afford fuel, woodlands are cut down, animal dung and crop residues are burned. Poor rural infrastructures constrain delivery of produce to markets and external resources to farms. Inequities in land tenure and other oppressions discourage resource conservation. Sustainable agriculture needs a favourable economic climate, with fair returns to farmers for labour, investment and the risks they bear.

Land Preparation and Protection

Before seeds or cuttings can be planted, soil has to be cultivated. The top soil is ploughed and turned over sometimes to depths of 30 cm or more, the lumps being broken to provide a granular seed bed. Deep ploughing of heavy compacted soils is demanding of mechanical energy and may exacerbate soil erosion. Where intensive production is common, there is encouragement for cultivation practices variously described as low-, reduced-, conservation-, conservative- or mulch-tillage, the objective being to minimise disturbance, degradation and erosion by leaving crop residues in the top soil. Low tillage improves drainage, water percolation and retention. Techniques include: use of chisel ploughs that penetrate to less than 15 cm but do not invert top soil; harrowing only sufficient to create a seed bed; seeding by direct drilling into stubble left from the previous harvest (CASAFA 1991).

The claimed advantages and disadvantages of any known tillage system are too many to be described, each being a reflection of a particular set of conditions. Deep ploughing can raise nutrients from sub-soils; control pest and pathogens by disturbing their habitats and by burying weed seeds. Deep ploughing may stimulate erosion by exposing bare soil to wind and rain.

Consequences related to low-tillage include:

- Soil nutrients and organic material tend to be concentrated in upper soils.
- Crop residues restrict water run-off with consequent loss of soil and nutrients. Fertilizer loss can be reduced by direct placement of coated granules.
- Weeds are left relatively undisturbed and may provide a favourable habitat for pests carried over from one season's crop to the next; this may impose greater dependence on herbicides and other pesticides.
- Residual mulch lowers soil temperature which can be beneficial in hot climates but a retardant on germination in cool environments.

Low tillage is generally better suited to moist than to very wet or very dry soils. Top soil in tropical drylands tends to be compacted and, after harvest, baked into a hard impenetrable crust that must be broken up before seeding. Time of tillage affects soil porosity, water percolation and retention, and nutrient distribution (Lal 1989).

No tillage system is perfect for all times and places. Low tillage with constant vegetative ground cover is protective against erosion, which many consider a first priority. For all agro-ecologies, research is needed to determine how tillage affects soil structure, nutrient and water distribution and retention; plant propagation and growth; natural soil organisms, beneficial and obnoxious; pests and their enemies.

Since most tillage practices have evolved empirically, change from an established to a novel system is best introduced gradually and with continuous monitoring to discover what changes in water management and other agronomic practices are best suited and impose least risk.

ATMOSPHERE, CLIMATE AND WEATHER

The Earth's atmosphere is a relatively thin blanket of air of which, though it extends far into space, more than one-half of its mass lies within 6 km of the Earth's surface. The Earth's atmosphere is unique within the solar system: whereas the outer planets' atmospheres consist mainly of hydrogen, helium or carbon dioxide, by volume this globe's atmosphere is composed of 78% nitrogen, 21% oxygen together with trace elements and compounds in various concentrations.

The atmosphere is a product of the biosphere, its natural composition is chemically unstable and, had plants not evolved, the oxygen content would be 1000 times less and the CO_2 content 100 times greater than they are (McBean et al. 1992; IGBP/ICSU 1992). Microbiological processes are largely responsible for raising the atmospheric concentrations of reactive gases, such as methane, nitrous oxide, ammonia and reduced sulphur compounds, gases with an important but only sparsely understood involvement in global climate control and the thickness of the Earth's protective ozone layer.

Solar radiation provides the photosynthetic energy by which plants consume carbon dioxide and release oxygen. Solar radiation, by uneven heating of the globe's surface, together with the planet's rotation and topography give rise to atmospheric and oceanic circulations.

The atmosphere's lowest layer, containing 80% of its total mass, is the troposphere, a system dominated by weather and water cycles. Water is critical to natural weather and climatic variations as it cycles through evaporation, transportation, transformation and precipitation. Water vapour in the troposphere is the dominant greenhouse gas.

Water sustains all plant and animal life, all natural and cultivated ecosystems, it replenishes reservoirs, rivers, lakes and aquifers. As it returns to the atmosphere as vapour it becomes the dominant greenhouse gas, and in the form of clouds it affects climate and weather patterns.

The stratosphere surrounding the troposphere has strong horizontal but weak vertical motions; pollutants that migrate to the stratosphere stay for very long periods. Layers of remarkable ionic complexity that extend outwards from the stratosphere, by their electromagnetic and gaseous structures shield the Earth from short ultraviolet radiations and solar wind particles from outer space. The thin ozone layer within the stratosphere plays a critical role in protecting life on Earth.

Weather is best understood in terms of temperature, rain, snow and wind, demonstrated in such extremes events as frost, floods, hurricanes, cyclones and drought. Climate, conventionally considered in 30-year norms, may be described as the statistics of weather. Climate is dynamic and varies over a wide range of time scales, influenced by oceans, ice-caps, and land surfaces, particularly whether bare or vegetated.

The Earth's mean surface temperature is determined by the intensity of solar radiation, by the albedo, the fraction of radiation that is reflected back into space, and the atmospheric thermal radiation that returns to Earth. Thermal radiation that returns is heavily influenced by atmospheric concentrations of "greenhouse gases", composed of water vapour, carbon dioxide, methane, nitrous oxide, ozone and, more recently, chlorofluorocarbons. Without greenhouse gases the Earth would have an intolerable mean temperature below -15 deg C (Golitsyn et al. 1992).

Climate and weather are influenced by natural phenomena and human activities. An ambitious international scientific effort is devoted to gaining a better understanding of what constitutes and controls global climate change. (ICSU 1992; ICSU/WMO 1990; IGBP/ICSU 1992).

The International Geosphere-Biosphere Programme (IGBP) studies those dynamics of the biosphere which influence and are influenced by global environmental changes. IGBPs objective is: "To describe and understand interactive physical, chemical and biological processes that regulate the total Earth system, the unique environment it provides for life, the changes occurring in the system, and the manner in which they are influenced by human activities." The purpose of the World Climate Research Programme (WCRP) is to gain a quantitative understanding of the physical climate system as constituted by four main components: Global atmosphere; World oceans; the Cryosphere; and Land surfaces. WCRP includes: (a) The Tropical ocean and global atmosphere study, with particular concern for the Southern Oscillation-El Niño phenomenon; (b) The Global energy and water cycle experiment; (c) The World ocean circulation experiment, with observations of all oceans to develop mathematical models of circulation and heat transport. (ICSU/WMO 1990).

Among many other published commentaries, the implications of global climate change for the future of agriculture and food security have been discussed by Parry et al. (1990) and CASAFA (1991). Intriguing to many authors are possible changes in stratospheric and tropospheric composition, the consequences for climate and weather, and in turn how agriculture and food production may be affected. There is persuasive evidence that destruction of the ozone layer will exacerbate the risk of skin cancer and depress yields of several food crops. Of ongoing yet unresolved debate is the question of how rising concentrations of greenhouse gases may affect the global climate and agricultural production. There is a body of opinion that rising gaseous emissions will elevate the Earth's surface temperature, alter weather patterns and precipitations. Rate and extent of climatic change will probably vary regionally, modifications being most marked across latitudes.

Parry et al. (1990) suggest that if greenhouse gas emissions persist at present rates global warming of *ca.* 0.5 deg C may be expected early in the 21st century, with a further rise of 3 deg C in the latter half of the century. Though exact effects on precipitation are uncertain, Parry et al. suggest that at extreme northern and southern latitudes rainfall could rise by 15% during winter months, while at mid-latitudes (30-45 deg. N and S) summer and winter rainfall will be lower than at present. Higher temperatures will raise evapotranspiration rates.

Atmospheric composition influences and is influenced by living organisms, including cultivated crops and livestock: methane emissions come from the flatus of ruminants and from rice paddies; nitrogen and ammonia emanate from decomposition of manure and chemical fertilizers.

High concentrations of carbon dioxide accelerate photosynthesis and increase vegetative biomass; in some species a doubling of CO_2 raises dry matter production by 15%. Plants may be classified according to biochemical pathways of photosynthesis and the mechanisms by which they assimilate carbon. Plants classed as C_3, that include wheat, barley and potatoes, respond productively to high CO_2 concentrations. C_4 crops, that include maize, sorghum, millet and sugar cane, are less affected, suggesting that raising atmospheric CO_2 concentrations will be of greater benefit to temperate countries that grow wheat and barley, than to poor semi-arid nations where sorghum and millets are cultivated.

Parry et al. (1990) discuss what might be the various agricultural implications and what adjustments in agricultural production practices may be necessary, and conclude that even with a considerable change in global climate, food production can be sustained sufficiently to meet world demand, though production levels within regions might undergo profound modification.

Atmosphere, Drylands and Deserts

Williams and Balling (1994) review various interactions between climate and dryland ecologies. They discuss "desertification", defined as: "land degraded in arid, semi-arid and dry sub-humid areas resulting from various... climatic and human activities". Roughly 20% of the world's population live in ecologies thus defined. Drylands are found where, annually, evapotranspiration (water vapour migrating into the atmosphere from plants and the earth's surface) exceeds precipitation.

Human beings affect soil and atmospheric composition in drylands by activities that lead to soil erosion,lower surface moisture retention, depletion of vegetation, particulate and gaseous emissions from the burning of biomass. Smoke emissions from biomass combustion in tropical developing countries, with consequent increase in atmospheric carbon and nitrogen compounds, exceed the mass of emissions from combustion of all fossil fuels. More than 25 gigatonnes (Gt) of smoke emissions come from the burning of close to 9 teratonnes of tropical biomass. Less than 24 Gt is emitted from all the world's fossil fuels, though the latter will escalate as more coal-burning industries in China and elsewhere come on stream. Roughly 43% of tropical biomass is burned in Africa: for firewood, in shifting cultivation, from stripping of residual forests, and by adventitious and deliberately ignited savanna fires.

In addition to the atmospheric pollution caused, burning of savanna woodland, grassland and forest exacerbates soil erosion by depleting vegetative cover,leaving land exposed to further degradation to be aggravated by wind and drought (Williams and Balling 1994).

The 120 nations that ratified the UN Convention on Climate Change, drafted in 1992 during the UNCED in Rio, undertook to take action, by the year 2000, to cap their "greenhouse gas" emissions at 1992 levels. At the time of writing relatively little progress towards the target proposed is evident. Jurisdictions in Europe and North America seem to have had second thoughts about imposing energy taxes to reduce fossil fuel consumption, in part because of the high cost of implementing fuel conservation and emission controls, in part because of the uncertainty of what will be the climatic consequences. If the cost to the affluent nations is considered excessive, the cost to developing nations, anxious to expand their manufacturing and transportation facilities, would likely be unbearable. China's energy consumption, two-thirds from coal, is forecast to rise five-fold in the next 25 years.

Present knowledge being incomplete, prophesies of future global and regional climate change in response to rising emissions of "greenhouse gases", though uncertain, are disconcerting. At a recent meeting of the American Association for the Advancement of Science, Dr. David Thompson of the AT&T Bell Laboratories reported how he applied a mathematical technique known as "Slepian tapers" to study the relation of rising carbon dioxide levels with seasonal cycles in the "anomalous year" - the time taken for the earth to orbit the sun, and the "equinoctial year" - the time between vernal equinoxes. Dr Thompson concludes that changes in carbon dioxide, resulting from human activities, are causing changes in average temperature and in seasonal climatic variations, particularly the date when winter starts in the northern hemisphere. From the analysis of data gathered in Britain since the 17th century, Thompson concludes that there is a strong likelihood of future climate change though the nature and magnitude of the changes expected appear not yet precisely determinable (Economist 1995).

Williams and Balling (1994) discuss various global climate models - "idealised representations of reality" - all of which predict future increase in earth surface air temperatures as atmospheric CO2 levels rise, but which vary substantially in the degree of temperature rise to be expected. Most such models also forecast higher levels of globally averaged precipitation. Though disparities and inconsistencies among models are apparent, greatest increases in precipitation are foreseen for the mid to higher global latitudes.

Williams and Balling (1994) discuss what they perceive as influences of the El Niño-Southern Oscillation phenomenon on global climate and rainfall patterns, commenting specifically on the Southern Oscillation Index: the surface barometric pressure difference between Tahiti and Darwin in Australia. Whenever the pressure in Tahiti is higher and the pressure in Darwin lower, the Index is negative. A negative Southern Oscillation Index has several times been associated with droughts in Sub-Saharan Africa and India, while a positive Index has often been accompanied by heavy rainfall, with floods in some Asian countries.

Williams and Balling point to the urgent need for standardised and generally accepted methods by which accurately to determine "desert spread", and to measure land degradation. Also needed are reliable early warning indicators of impending drought, with people who are trained to monitor early warning indicators and give timely alert to governments, farmers, development and relief agencies.

Williams and Balling (1994) present recommendations for research and other actions needed to restrict future serious land degradation and for drought alleviation in African dryland ecologies. It is in these stressed ecologies that the world's poorest nations struggle to exist.

Drought

Hulse and Escott (1986) discuss drought in historical perspective, the contributory causes and consequences, the regions most afflicted, and remedial measures needed.

Derived from an Anglo-Saxon word meaning dry land, "Drought" is a comparative not an exact term. Meteorological drought describes total rainfall below long-term average over an extended area and period of time. Agricultural drought indicates conditions resulting in low crop yields from a combination of low soil moisture with evaporation rates significantly higher than precipitation. For rain-fed crops, the time and duration of precipitation is as critical as the amount precipitated

Food crops are hygrophytes: they need water throughout their growth cycles. Water deprivation after sowing restricts germination; after germination it causes desiccation of plantlets before they flower and fruit. Too little water during floral development and pollination may reduce the number of seeds formed, and after pollination result in smaller, shrivelled seeds. Low soil moisture inhibits uptake of soluble nutrients through roots. Insufficient water restricts photosynthesis and translocation of photosynthates from leaves to seeds.

In dryland areas, rain storms are often localised, as much as 20 mm/h falling at the storm centre, close to zero 1 to 2 km away. Exactly when and where drought will strike is unpredictable, but the African regions most frequently afflicted have been known for many centuries. Famines recorded in Egypt and northern Sudan resulted from failure of summer rains in the Ethiopian highlands. Ancient Egyptians 5000 years ago stored grain produced in surplus during years of plentiful rains, as strategic reserve stocks to be drawn on in times of drought.

Over 20 of Africa's poorest nations suffer recurrent drought. These are countries whose average GNP/person range from $80 to $650, life expectancy from 38 to 57 years and where the population is forecast to increase to 350 M by the end of the century. The ravages of droughts in Sahelien and southern African countries have been well publicised. But once the dismal pictures disappear from television screens the threat of droughts to come is soon forgotten. The need is urgent for research to discover the biochemical and physiological mechanisms whereby crops evolve tolerance to drought stress. Equally necessary are strategic reserve stocks to be carried over from regions and seasons of surplus to those of drought-imposed scarcity (Hulse and Escott 1986).

There exists an impressive body of scientific knowledge on food crop resistance to drought stress, knowledge enlarged by research in Asia, Africa and Latin America, and in Australia, Canada and the USA all of which experience severe droughts roughly every 20 years. Urgently needed is an international cooperative program to integrate the various independent programs, international, regional and national, now in progress (Baker/CASAFA 1989).

The Desert Margins Initiative (DMI), derived out of the International Convention on Desertification (INCD), offers hope that cooperation among international and national research centres may, in time, retard and reverse land degradation in dryland, drought-stressed zones of Africa. DMI brings together the resources of four International Agricultural Research Centres with several African regional and national research centres and programmes. It also purports to include, cooperatively, the practical experience of several non-governmental organisations (NGOS) who have maintained agricultural programs in drought-stressed African countries. The DMI was launched at a meeting in Nairobi in January 1995 and will be sponsored and supported by the CGIAR consortium of donor and development agencies.It will study past experience and bring about new collaborative research relevant to land degradation. It will include crop and soil sciences, livestock and range management, together with agro-forestry and agro-meteorology (ICRISAT 1995).

LAND AND WATER MANAGEMENT

Soil Conservation

The formal literature on soil fertility, conservation and degradation is vast. A bibliography is published in CASAFA 1991. Other references include IISAA 1990, ISRIC 1990, WRI 1992, Swaminathan and Sinha 1986, AIC 1994, Lal 1989 and for the USA NRC/NAS 1989, USDA 1986, 1987, 1988.

At time of harvest the dry matter of a grain crop consists mainly of carbon, oxygen, hydrogen with about a dozen other elements, most derived as minerals from the earth's crust, from which during weathering top soils and subsoils are formed. In the weathering process nutrients are released to be taken up by plant root systems (van Dienst 1991).

The minerals that constitute the earth's crust are widely diverse in chemical composition, the rate of release during weathering is slow, so that nutrients present in top soils are often insufficient fully to meet the nutritional requirements of cultivated crops. Nutrients in which natural soil is deficient must be supplied from external sources such as manure, compost, crop residues and synthetic fertilizers.

Above each hectare of the earth's surface roughly 75,000 t of the element nitrogen float in the atmosphere. In theory, nitrogen need never be in short supply. Elemental atmospheric nitrogen cannot be used directly by growing plants but must be converted by microorganisms to ammonia or some equally assimilable compound. To combine nitrogen with hydrogen, by natural biological processes or by chemical synthesis, demands significant energy. Symbiotic nitrogen fixation in the root nodule of a legume demands the bond energy from *ca.* 15 moles of ATP for each mole of ammonia formed. Conversion to ammonia of nitrogen and hydrogen in a Haber-Bosch reactor takes place at 500 deg C under a pressure of 20 million Newtons per sq metre.

Essential Soil Nutrients

Van Dienst (1991) describes an agricultural system as "sustainable" when no more nitrogen is removed in harvested produce than enters the soil by the action of nitrogen-fixing microbes. Production of gaseous hydrogen and its chemical combination with nitrogen is expensive, and so it is desirable for poor countries with limited energy resources to develop agricultural systems that are not wholly dependent on imported nitrogen fertilizers.

Symbiotic nitrogen fixation in a legume root nodule imposes a high demand for phosphorus. Rock phosphate deposits are found in several developing regions but not all food crops can directly use ground rock phosphate. The International Fertilizer Development Centre in the USA, and its affiliate in Africa has developed simple means of utilising rock phosphate in cropping systems (IDRC 1981).

The demand for potassium, an element essential to crop cultivation, is 10 times higher than the need for phosphorus in many plants. Natural deposits of potassium are more abundant in affluent northern countries than in tropical or subtropical regions where it is relatively scarce. Concentrations of potassium in ocean waters is *ca.* 5000 times that in terrestrial deposits. Technologies for the separation of potassium from sodium in sea water are available but demand high energy inputs.

The 10 micro-mineral nutrients essential to most plants generally carry fewer limitations than nitrogen, phosphorus and potassium, though distribution and availability varies among different soils.

Van Dienst (1991) discusses alternative options open to resource-poor farmers to reduce dependence on expensive agro-chemicals. These include: nutrient recycling, mixed farming and crop rotations.

Nitrogen, phosphorus and potassium are essential to every crop's organic processes, the amounts and times of greatest need varying among species, stages of reproduction and growth. Chemical fertilizers deliver nutrients when a crop most needs them and in forms designed for easy uptake. Analyses relate soil composition to cultivated crop needs. Agro-chemicals provide deficient nutrients and neutralise substances detrimental to plant growth. Soil acidity can be controlled by lime at levels below which elements toxic to plants become soluble.

Nutrients removed from soil by growing crops can be restored from various sources: animal and green manure, crop residues, leaf mulch and chemical fertilizers. Green manures derive from terrestrial or aquatic plants cultivated and/or harvested to be used as organic sources of nitrogen and other nutrients. They may stimulate biological activity and restrict erosion. Particularly useful are multi-purpose trees, their various parts being usable as fertilizer, forage and fuel (Huxley 1983).

Maintenance of soil fertility is essential to sustain production. Nutrients can be lost from soil by wind and water erosion, by leaching and reactions that reduce their solubilities. In traditional shifting cultivation, crops are planted and harvested until the soil nutrients are exhausted. During long regenerative periods, trees and other deep rooted plants suck up nutrients from sub-soils and feed them to top soil as leaf litter decomposed by microorganisms. Increasing population densities have caused shifting cultivators to shorten the regenerative phase leading to serious soil degradation in many tropical regions.

Over the past 25 years, high yielding varieties that respond well to chemical fertilizers, have raised Asian rice harvests by over 60%. Excessive use of agro-chemicals can impair organic matter content, and various remedial measures are proposed (Swaminathan and Sinha 1986; van Dienst 1991).

Alternatives to Agro-Chemicals

Organic sources proposed include animal and green manure, composted residues and leaf litter, and more extensive cultivation of legumes. Animal manure is rarely used very efficiently. Manure from cattle owned by migrant pastoralists largely dries and blows away. Kerosene and wood being too expensive or inaccessible, manure is often burned as fuel. Nitrogen is released faster from manure than from crop residues high in lignin. Rates of release, varying with soil temperature and moisture, by season and location, are not easily controlled. Nutrient loss from manure collected and stored on farms, results from liquid run-off, from leaching and volatilisation of ammonia. Reported nitrogen loss from various forms of solid and liquid storage range from 15 to 80%. Other losses vary with methods of distribution and application. Methane generated by anaerobic fermentation of manure to be used as biogas, seems of uncertain and variable economic viability (ERG 1986).

From heavy weathering, leaching, intensive cultivation and stressed by high population densities of very poor people, tropical soils generally are low in essential nutrients, and in greater need of supplementation than soils in temperate zones.

Legumes, which grow pod-shaped fruits, are less dependent upon soluble soil nitrates since *Rhizobia* bacteria resident in root nodules can fix and convert atmospheric nitrogen to a usable form. Unharvested legumes mulched or ploughed-in can provide nitrogen usable by other crops, but harvested legumes may remove more soil nitrogen than they fix. Studies of legumes grown at different locations report wide variations in the levels of nitrogen fixed (NRC/NAS 1989). The following (kg/ha/year) illustrate the range of variability in symbiotic nitrogen fixation:

Alfalfa	78-220	Birdsfoot trefoil	50-110
Phaseolus beans	2-215	Faba beans	117-250
Field peas	174-195	Lentils	167-190
Red clover	68-115	Soybeans	22-310

Though more than 200 legume species are known to be cultivated, their potential as contributors to soil nitrogen is far from being realised. In addition to *Rhizobia*, which fix nitrogen in symbiotic association with legumes, many soils are home to free-living microorganisms that fix atmospheric nitrogen. Among them are temperate and tropical species of *Azotobacter* and *Spirillum*, tropical species being more efficient. Biological nitrogen fixation in tropical ecosystems is comprehensively reviewed by Mulongoy et al. (1990).

Other microorganisms provide nutrients by decomposing organic macromolecules present in leaf litter, crop residues, compost and manure. Some synthesise carbohydrates that bind soil particles together. Species of mycorrhizal fungi, about which much remains to be learned, stimulate phosphorus uptake, root proliferation and seedling vigour. Though increasing use of chemical fertilizers may cause legitimate concern among ecologists, there is no convincing evidence that, when used prudently, chemical fertilizers pose any greater threat to the environment or to human health than alternative sources of nitrogen, phosphorus, potassium and other nutrients. Optimum times of application rates of release and nutrient availability are more predictable from agro-chemicals than from most organic sources. For chemical fertilizers to be superseded, without severe loss in productivity, much research is needed on the complex relations and interactions that control nutrient synthesis, storage, cycling and availability among diverse cropping systems, in soils widely different in structure and composition.

Soil Degradation

Soil is a living system: its essential structure, composition and biological diversity must be understood and protected for productive agriculture to be sustainable. Though the principal causes of soil degradation can be described, quantification and future prediction are difficult particularly in fragile ecologies (Brown and Wolf 1984).

Arable land varies in area and quality among and within continents, the global average being estimated at about 0.5 ha per living person. Land presently cultivated varies from less than 0.2 ha per person in Asia and Western Europe to about 0.33 ha in Sub-Saharan Africa and >1.0 ha in North America. Cultivable land is roughly 1.5 ha per person in Africa, *ca.* 1.9 ha/person in North America.

The proportion of available arable land already cultivated varies from *ca.* 85% in Asia and Western Europe, to *ca.* 50% in North America, 22% in Sub-Saharan Africa, and 11% in South America. Land classed as arable in Africa includes large areas of tropical forest, and humid zones larger than the USA that are heavily infested with tsetse fly, a vector for sleeping sickness in humans and severe anaemia in cattle. Tropical forest soils tend to be fragile and rapidly eroded when tree cover is removed (CASAFA 1991).

Soil degradation includes decline in physical structure; erosion by wind and water; depletion of organic matter and natural nutrients; toxicity from salinity, alkalinity, acidity and chemical contaminants. Erosion occurs under natural conditions but is aggravated by careless human interventions. Soil may appear infinitely auto-regenerative where conservative production systems have long been maintained, but as dependent populations increase, and higher crop yields are sought from limited arable land, soils are overstressed and degraded. Trees, shrubs and grass bind and protect soils and prevent run-off letting rainwater percolate downwards to recharge aquifers (NRC/NAS 1989; Huxley 1983).

Production potential: the quantity of plant material (phytomass) that a particular soil can sustain varies from near zero in Arctic tundra and arid tropics to an estimated 450 t/ha in the humid tropics. Within any ecology, production potential depends upon the many factors and resources that control plant growth, soil fertility and conservation. When more is removed than is restored, production potential goes down.

Overlooking pests, pathogens and predators, natural constraints to productivity include drought (28), inadequate soil structure, composition and fertility (23), shallow depth of top soil (22), excessive water (10) and permafrost (6), percentages of the earth's land surface so affected being shown in parentheses (CASAFA 1991).

A UN Environmental Agency report estimated that over 15 million ha of agricultural land are destroyed or degraded annually: 8 M ha by conversion to non-agricultural use; 3 M ha by wind and water erosion; 2 M ha by toxicity; 2 M ha by deterioration to desert. Non-agricultural includes conversion to industrial and urban sites; roads, railways and airports; garbage dumps and sewage treatment plants; mining and mineral extraction (UNEP 1982). Egyptian scientists have demonstrated that deserts can be reclaimed for agriculture (El Lakany 1991). Soil buried under asphalt, bricks and concrete is lost forever.

The causes of degradation are many: over-cultivation, over-grazing; removal of ground cover and trees that act as wind-barriers; excessive tillage especially of very wet and very dry soils; continuous strip monoculture of shallow-rooted crops; excessive or poorly controlled irrigation; soils compacted by heavy equipment or trampling by animals in high stocking densities (compacted soils impede water percolation and retention); chemical contamination by industrial, urban and vehicle emissions, acid rain, accumulation of metal and pesticide residues (CASAFA 1991).

Tropical Soils

Degradation is severe in fragile, shallow, heavily weathered, leached tropical soils, nutrients, often imperfectly balanced, being concentrated in thin top layers. Shallow compacted soils in the semi-arid tropics tend to be low in organic matter, prone to crusting and resistant to cultivation and water penetration (Greenland and Lal 1977; Norse et al. 1991) In India, erosion degrades about 60% of crop land resulting in an annual top soil loss of over 4.5 billion tonnes.

Erosion is most severe on hillsides and slopes from which ground cover is removed. In an East African country, on adjacent land similar in degree of slope, soil covered by bush and grass suffered no erosion; under continuous monocrop, annual soil erosion was 64 t/ha; under bare fallow 125t/ha. From the soil covered by bush, water run-off was less than 4%, from fallow over 50% of precipitation. In another African location on tilled land, from a 1% slope top soil loss was 1.2 t/ha, from a 15% slope 24 t/ha. Erosion from slopes can be reduced by terracing, contour cultivation and cropping systems that maintain permanent vegetative cover (CASAFA 1991).

Detrimental effects of erosion extend far beyond farms. Eroded soils carried into surface waters cause silting and pollution. Organic matter deposited in lakes and rivers deprives aquatic species of essential oxygen. Silting seriously reduces the effective purpose of reservoirs constructed to provide irrigation or hydro-electric power.

Soil Erosion in North America

During the 1930's soil erosion in North America was exacerbated by drought in the U.S. Mid-west and Canadian prairies. After WWII highly mechanised farm production systems dependent on agro-chemicals rapidly expanded. Increased crop yields were accompanied by impaired soil quality and encroachment upon the environment (NRC/NAS 1989).

In Canada, soil quality has suffered as the result of: (a) regional specialisation with crop cultivation widely separated from livestock husbandry, reduced use of forage crops and animal manure; (b) reliance on summer fallow that exposes soils to erosion and loss of organic matter; (c) bigger farms using heavy machinery causing soil compaction, perimeter trees cut down to enlarge fields. Canadians estimate that water erosion has affected 5.8 M ha, wind erosion 6.4 M ha, salinity 2 M ha and compacting 1 M ha, the annual cost being assessed at close to $950 M. Organic matter in many prairie soils is less than half of the original content (AIC 1986; Senate of Canada 1984).

Reports from the USA suggest that one-third of the topsoil has been eroded from cultivated land over the past century; that 25% of crop land is seriously affected; that *ca.* 3 billion tonnes are eroded annually; that salinity is of increasing severity on irrigated land; that over 50 M ha are prone to erosion. Main causes are continuous row mono-cropping, few crop rotations that include forage legumes, ever larger farms using deep tillage and heavy machines (NRC/NAS 1989).

The U.S. NRC states that, after 50 years of state and federal efforts to control it soil erosion remains a serious environmental problem . Erosion causes off-farm and on-farm damage, off-farm costs estimated at between $2 billion and $8B annually. The U.S. Department of Agriculture estimates that 150 million ha of agricultural land account for more than 50% of suspended sediments deposited yearly in surface waters. (USDA 1986, 1987, 1988). Different estimates of the annual cost of on-farm erosion vary from $1 billion to $18B, the wide disparity being largely from differences in criteria and methodology (Sampson 1984).

The NRC report estimates that some 40 million ha of farm land now employ some form of conservative tillage, an area that has more than doubled since 1980.

Sampson (1984) predicted that in the next 50 years less productivity caused by soil erosion of U.S. cropland will be equivalent to the loss of more than 10 million ha, an area that could yield over 50 million tonnes of grain, roughly half the quantity exported in 1980. Sampson also states that between 1967 and 1977, 2.5 M ha of U.S. cropland were converted to non-farm use, which in economic terms would produce about 2 Mt of maize worth over $200 M. At least 40% of the U.S. land converted to non-farm use was first class land. Sampson contends that from the southern tip of California through southern Nevada, Arizona and New Mexico into the western half of Texas, also in parts of Oklahoma and Kansas, desertification, affecting over 90 M ha, is spreading because of progressive soil degradation, loss of moisture retention and vegetative cover. Soil salinity, from mis-managed irrigation, further restricts crop production. Sampson advocates a shift from chemical fertilizers to nutrient recycling but accepts it may be several years before crop yields can be sustained at levels to which farmers are accustomed.

There is general agreement on measures that would lead to reduced degradation in North America: multiple cropping with rotations that include legumes; reduced tillage; maintenance of vegetative ground cover with trees as wind-breaks; mixed crop and livestock systems; nutrients from animal and green manure; less dependence on chemical fertilisers.

Reasons given for their not being widely adopted include: too little research on which to base low-risk, economically, ecologically sound production systems. In absence of credible research and demonstration, benefits and risks to farmers disposed to adopt conservative practices are too uncertain. Farmers with heavy incurred debt and high production costs are reluctant to change from known systems that yield good short-term cash flows, to production practices the consequences and returns from which are unpredictable (Sampson 1984).

Resistance to Remedies of Soil Degradation

As with all other issues that bear upon agricultural production systems and how they may be economically sustained, there are no simple courses of action applicable to all conditions. In international and most national agricultural research programs, more is invested in crop species improvement to increase yield potential, pathogen and pest resistance, than to soil biology, chemistry and physics; to soil erosion, protection and conservation, and to related economics.

In recent years some, though relatively few government legislators have passed laws to prohibit the degradation and destruction of arable land by urban and industrial spread and for other non-agricultural purposes (AIC 1986). Since, globally, arable land is much smaller in area than land that cannot be cultivated, it would seem sensible to use poorer land for urban and industrial habitation, and to protect and preserve cultivable land for agriculture. Unfortunately, areas with the most fertile soils and favourable climates are the pleasantest places to live and work. Industrial and urban developers reap higher returns from building over good land than do farmers from cultivating it. As farm land is destroyed by bulldozers, bricklayers and cement mixers, the farm land remaining will inevitably be subjected to more intensive and debilitating production systems.

It is a tragedy that every year as 90 million more mouths arrive on earth to be fed, the area of arable land rendered unfit for cultivation could, at average global yields, provide cereal crops sufficient for at least 90 million people (CASAFA 1991).

Water

The Earth's fresh water resources are seriously polluted from human, municipal, agricultural and industrial waste. Two-thirds of global water is used for agriculture, roughly a quarter by industry. Early in the next century industrial withdrawals will more than double (WRI 1992).

Water is the planet's most abundant resource, covering close to 70% of the globe's surface. Were it spread evenly the water would form a layer roughly 3000 m deep. As it exists, almost 98% of the planet's water lies in the oceans and inland seas and is too saline to be potable or usable for crop irrigation. Of the 3% free of salinity, most is locked up in ice caps or glaciers, as atmospheric vapour, and in aquifers deep below ground. Useful sources of fresh water exist in rivers, reservoirs and lakes, almost one-half in South America, a quarter in Asia. Fresh water is replenished by precipitation: about 5000 km^3/year, only one-fifth of which falls on land, of which close to 65% evaporates into the atmosphere. What remains flows to reservoirs, rivers, lakes and wetlands, or percolates down into aquifers (Stewart et al. 1992).

Levels of precipitation vary extensively across the globe, heavy rainfall being climatically characteristic of the Amazon basin and the monsoonal regions of Asia, while over 80 countries home to 40% of the population and 30% of cropland suffer serious periodic drought.

Human use of water is rising at over 5% per year, fastest growth in demand being in developing countries. Annual per capita use varies from 1700 cu m in North America, to less than 250 cu m in Africa. Roughly 85% of water withdrawn in Asia is for crop irrigation. While throughout the world demands by industry are rising rapidly, the need for irrigation will also continue upwards as the need for larger crop harvests are driven by burgeoning populations in need of more food.

Water in North America

Pollution of fresh water is of serious concern as lakes and rivers are contaminated by soil erosion, agricultural and industrial chemical effluents, pathogens from untreated sewage, heavy metals and noxious compounds from mining and forest products industries (WRI 1992).

U.S. Agriculture consumes close to 85% of total water use. The U.S. Water Resources Council predicts water consumption will rise with greater competition to agriculture from industries and power generation. WRC also anticipates, early in the next century, large areas of the mid-west and south-west to suffer severe depletion as groundwater aquifers are run down. Crop irrigation is drawing from groundwater aquifers faster than they are being replenished (WRC 1978). Unless water is used more conservatively, crop irrigation will be forcibly reduced. Californian farmers are charged less than 10% the cost of the water they use.

English (1984) and LeVeen (1984) present critical assessments of water in relation to food security in the USA. Relative to food security, water availability must be studied as it relates to other essential resources.

Consistently rising crop yields in the USA have ensured that over the past 50 years production has substantially exceeded local demand. In the 1960's 25 M ha of crop land were withdrawn from production. Rising oil prices, together with severe drought, followed by more productive years, led to wide fluctuations in grain prices during the 70's and 80's. After the 1973 World Food Congress predicted serious food shortages, U.S. farmers were encouraged to produce to the limit of their capacity. Their response led to a doubling of wheat exports. Irrigation water came from wells sunk in the Great Plains to bring previously unproductive land under the plough.

The U.S. Water Resources Council estimates: on average 160 B hL of water falls as precipitation every day, about two-thirds of which rapidly evaporates or is transpired to the atmosphere by vegetation. Disappearance via rivers into oceans, later evaporation and other loss reduces the daily U.S. water supply to *ca.* 25 B hL (WRC 1978).

Estimated daily consumption is *ca.* 15 B hL, indicating that projected demand falls far short of anticipated supply. Precipitation, surface and ground water are unevenly distributed across the USA. The Water Resources Council states that whereas New England, Ohio, Tennessee and the Upper Mississippi are adequately supplied, large areas of the lower Colorado, California, Texas Gulf and Arkansas heavily overdraw their ground and surface waters. In these drier areas water is being drawn from aquifers much faster than it is replaced. As water tables fall, the energy costs of pumping will continue to rise until conservation restores an effective balance. USWRC estimates that practicable methods of conservation could save over 600 B hL/yr (WRC 1978).

Economies may be realised by lining or covering canals, better land preparation, scheduling water release at times of least evaporation (e.g., at night) and more efficient uptake: on-farm irrigation efficiency ranges between 10 and 40%. Conservation up-stream of farms needs to be improved by harvesting water before run-off and reducing loss between reservoirs, rivers and farms.

Water conservation is hampered by a lack of systematic records and data from past experience. The Centre for Agricultural and Rural Development (CARD) includes ground and surface water data in its programming models that examine demand levels, water prices, alternative crop land use. At present the models are more suggestive of what should take place than predictive of what will occur as circumstances change: as availability falls and as marginal prices vary.

English (1984) contends that agricultural land is not the resource most limiting to increased U.S. production. Existing technologies can more than satisfy demand for the foreseeable future. About 50 M ha of range and forest land could be converted to crop cultivation; wheat yields could be raised by 50%, soybeans 60%, maize 40%. Livestock and dairy output could increase by 25%.

Prices paid for water by farmers is far below true market value. It is difficult to predict how retail food prices and commodity export prices would rise if farmers were charged a true market value for water. Continued subsidisation of water to U.S. farmers forebodes a relentless depletion of ground and surface water in dry States where aquifers are already seriously overdrawn.

LeVeen raises the prospect of the 70's energy crisis being superseded by a global water crisis. Because of uneven distribution, available water in the U.S. varies among States and years. Water conservation in the U.S. is urgently needed but ill-designed reservoirs and delivery systems can cause ecological disruption. Capital and maintenance costs to conserve and use irrigation water more efficiently will raise farm commodity prices faster than other goods and services (LeVeen 1984).

WRC (1978) has recorded total water use as a percentage of stream flow for *ca.* 60 U.S. river systems. In years of average precipitation, from 28 of the 60, total water use exceeds stream flow. During dry years, in only five is stream flow higher than total use. Excessive draw-off is most evident in southern warm, dry states. Agriculture consumes over 85% of total water used in the western states where there is competition from growing urban communities and industries which can afford higher water prices than farmers. Water supply will be evermore politically sensitive as competition intensifies and costs of delivery rise.

To move water over long distances is expensive, energy intensive and technically difficult. Quantities to be captured by dams, canals and aquifers are small relative to rising demand and would incur higher capital and maintenance costs than would attract private financiers. It is estimated that each 100 hL of new water in California would cost between 80¢ and $3.20. California farmers pay less than 10¢ per hL. Given the difficulty of predicting how crop prices would respond to higher charges to farmers, cost-benefit analyses of investment in new water delivery systems are quite elusive. Proposals, such as those to divert the Yukon and Mississippi Rivers to transport water over huge distances to the western states would be exceptionally expensive and technically complex, and likely to arouse opposition from environmental activists. Debt-stressed farmers subjected to higher water prices might be induced to sell their farms or water rights to non-agricultural users, with unforeseeable consequences for U.S. crop production.

Eventually, in absence of any remarkable technological innovation, costs of energy and water distribution will rise. As discussed earlier, measures to conserve and use irrigation water more efficiently all incur costs. Past experience indicates that governments are generally more favourably disposed to investment in new innovations than in conservation. As costs of delivering water to the drier states escalate the government may legislate to encourage grain crop production where water is more plentiful. In the warmer drier states irrigation may have to be restricted to horticultural crops of higher value.

Crop Irrigation

Of 150 M ha of U.S. farm land in crop production, barely 21 M ha (*ca.* 14%) is irrigated, of which 18 M ha are in western low rainfall areas. Statistics of irrigated crop production published by USDA are tabulated below.

	Area irrigated (% total)	Production ratio Irrig.:Non-Irrig.
Wheat	3.9	1.76:1
Maize	8.5	1.27:1
Sorghum	40.8	1.86:1
Barley	15.5	1.73:1
Dry beans	35.0	1.58:1
Cotton	29.0	1.90:1
Hay	18.0	1.64.1

Given water delivery costs, it would appear as sound economic sense to cease irrigation of cereal grains, feed and forage crops in states where water resources are low and being fast depleted.

Of 3.5 M ha of irrigated land in California less than 1 M ha is planted to fruits and vegetables. If Californian farmers were charged the full market cost of water it is likely that many would replace irrigated rice and pastures with high value horticultural crops. Barely 10% of U.S. cropland grows cereals, fruits and vegetables for direct consumption by Americans. Over 40% of cropland and 30% of farm water is given to the production of export crops, 65% of which are purchased by other OECD and eastern European nations. A sizable proportion of irrigated grain sold to LDCs is fed to livestock (LeVeen 1984).

Final Comment on Water

The belief appears widespread across the planet that water, in common with such other resources as natural forests, belong to no one in particular, therefore they belong to everyone and are free to be taken, wasted and despoiled. Water management in irrigated agriculture, is not an exact science. Nonetheless, if profligate waste and severe ecological disturbances are to be avoided, substantial investment in soil and water research is urgently needed, better to understand and control the complex relations among water, soil and plants in the many and diverse agro-ecological locations and conditions around the world where crops are cultivated.

ENERGY IN AGRICULTURE

Following the energy shock of the 1970's, when the OPEC countries tripled the price of oil in 1973, the industrialised western nations rapidly improved the efficiency with which energy is used, domestically and industrially. The World Resources Institute calculates that among developing countries consumption of commercial energy has almost tripled since 1970, coal and oil being the major sources. An additional 3-fold increase in demand is expected between 1985 and 2025, fossil fuels continuing to be the main sources (WRI 1992). Global commercial energy production rose by 52% between 1970 and 1990, oil and coal being respectively first and second in level of demand. Primary electricity from nuclear, hydro, geothermal and wind provide less than 5% of total. Since 1980 energy consumed in Africa and Asia rose by more than 50%. Tables 29, 30, 31 give patterns of growth and sources; Table 32 compares household consumption with rising expenditure; Table 33 estimates the effects on global warming of greenhouse gases from substances derived from energy generation and other commercial activities.

Table 29. Energy consumption.

	Total Consumption (MTOE)		Per Capita (TOE)	
Year	World	LDCs	OECD	LDCs
1970	15	<2	5.3	0.04
2000	30	12	6.0	0.45
2030	53	30	5.4	1.80

Total proven reserves: Oil and Gas 250 BTOE (TOE = Tonnes Oil Equivalent; B = Billions; M = Millions). One TOE approx equivalent to 45 GJ.

Table 30. Energy from different sources.

	OECD %	Developing %
Oil	37	23
Coal	25	28
Natural Gas	23	7
Nuclear	5	1
Hydro-Power	6	6
Biomass	3	35

Slessor (1986) reviews the balance between the energy consumed and produced by agriculture in the industrialised regions. Until the 1960's, energy policy was largely confined to considerations of which sources carried least marginal cost. Little attention was given to energy used in chemical fertilizers. Slessor refers to an original study reported from Strathclyde University which demonstrated that the energy needed to produce one kilogram of fertilizer had dropped from over 300 MJ early in the century to 17.5 MJ by 1970.

Table 31. Commercial energy consumption.

	Increase: Percent/year		
	1960-70	1970-80	1980-91
China	4.2	7.5	7.3
India	5.0	6.4	7.2
Malaysia	11.4	5.4	7.9
Thailand	16.2	7.9	7.4
Indonesia	3.7	9.9	4.8
Philippines	8.4	4.0	1.9
South Korea	13.9	10.1	7.9
Chile	6.1	1.0	3.5
Mexico	7.7	8.7	1.4
Venezuela	7.1	4.4	2.3

Commercial forms of primary energy only. Wood, organic residues excluded.

Table 32. Household consumption.

	Percent of Expenditure	
	Food	Fuel & Power
Low income nations GNP/Cap <$650	40-65	1-7
Low middle income $650-$2,500	23-51	2-15
Upper middle income $2,500-$10,000	19-35	1-7

Slessor defines "Energy ratio"(ER) as the quotient of the output of metabolisable energy divided by the input of fossil/fissile energy. He states that between 1880 and 1976 the ER in German agriculture fell from 4.0 to 2.0 as mechanisation and fertilizer use increased.

Table 33. Sectors: percent to global warming.

	CO_2	CFC	C_2H_4	O_3	NO_x	Total
Energy	35	-	4	6	4	49
Deforestation	10	-	4	-	-	14
Agriculture	3	-	8	-	2	13
Industry	2	20	-	2	-	2
Total	50	20	16	8	6	100

Percent increases over past 20 years			
World	USA/Canada[a]	OECD(Europe)[a]	PR of China
32	20	4	140

[a]OECD nations generate 36% of all greenhouse gases.

Energy Use and Efficiency in Agriculture

It seems conventional wisdom that a low energy ratio is to be condemned, a high ratio applauded. However, had chemical fertilizers not been discovered and mechanisation of production and post-production systems left undeveloped, Malthus' melancholy prophecies might have been fulfilled. Slessor presents Gross Energy Requirement (GER), energy (MJ/kg) to produce a unit weight of crop, Output Intensity (yield kg/ha) and Energy Ratio for various crops grown in different countries. Slessor's data are given in Table 34; they illustrate the energy input needed to raise crop yields in the USA and U.K. Slessor writes that while ERs of production systems do not vary greatly among countries at comparable Output Intensities, ERs vary significantly with production intensities.

Energy demand for production, preservation, processing, distribution and utilisation of food commodities will continue to rise well into the next century. Expanding urban populations, rising incomes, decline and deterioration of arable land per capita, collectively portend more intensive production systems with concomitant higher energy demand. Government control of food prices presses farmers to produce with greater economic efficiency and to raise crop yields. If all mankind became vegetarian agricultural energy demand would be less extravagant but, for reasons given, demand for animal products among the prospering nations of Asia and Latin America will inexorably increase.

Lavoisier had reason in writing: "La vie est une fonction chimique" since every organic process rests on the chemical conversion of one form of energy to another. Sinha (1986) quotes De Wit as defining agriculture as a human activity that converts fossil fuel into food. The Technical Advisory Committee of the CGIAR relates agro-chemical fertilizers to "using oil to save soil" (TAC/CGIAR 1989).

Table 34.

Crop	Country	Yield (t/ha)	Energy Ratio	GER (GJ/t)
Wheat	W Canada	1.5	8.7	1.6
Wheat	USA	2.1	2.4	5.7
Maize	U.K.	5.0	2.3	5.3
Maize	Mexico	1.9	10.7	1.4

Energy in Developing Country Agriculture

Sinha (1986) comments on the rising demand for commercial energy by agriculture in developing countries and suggests: (1) greater reliance on renewable sources; (2) improved efficiency of energy use, both of which call for critical analyses of energy consumption by different food and agricultural systems. Sinha reviews various energy sources and comments that fossil fuels and hydro-electric power which carry a cost must be treated as agricultural inputs. Solar energy being cost-free may be disregarded as an input. Among industrialised countries, human energy in agriculture is small enough to be ignored, whereas in poorer developing countries human and animal power are important factors.

Sinha (1986) compares energy consumption among agricultural systems and suggests how efficiency could be improved: (1) more reliance on biological nitrogen fixation than on synthetic fertilizers; (2) improved irrigation management; (3) mixed farming to replace mono-cropping; (4) energy farming to generate renewable plant biomass for fuel and power generation.

Photosynthetic Efficiency

Wittwer (1986) advocates research to raise photosynthetic efficiency in food crops. For most major food crops efficiency of solar absorption and utilisation during the growing season is barely 1%. Under favourable conditions, C_4 plants, that include sugar cane, maize, sorghum and millets, are photosynthetically more efficient than C_3 plants. Wittmer suggests, to enhance photosynthesis, research to identify the mechanisms that regulate light and dark induced respiration. Wittmer proposes genetic modification of food crops to increase the Harvest Index, those parts of cultivated plants which are of economic value.

Energy in Grain Production

Pimental (1984) quotes from Pimental and Burgess (1980) who compared energy output:input in three different maize production systems: (1) Hand-powered; (2) Tractor-powered; (3) Horse-powered. Primitive handpowered practised in Mexico requires one person with an axe, hoe and seed. The estimated output: input is *ca.* 12.5. Replacing hand labour by a tractor reduced total time for essential operations significantly but achieved an output:input ratio of 2.9. Using horse power, operational time was greater than for tractor power but less than for human labour, with an output:input ratio of 3.4.

Pimental et al. (1980) estimated that 90% of grain used in the USA (*ca.* 135 mt) is fed to livestock. The fossil energy absorbed to generate one kcal of animal protein in the USA varies widely among husbanded species and how they are fed. The following are indicative of kilocalories needed to produce 1 kcal of protein:

Dairy cattle (grass fed)	7	Hogs	43
Broilers	10	Sheep	88
Eggs	18	Beef cattle	25

Pimental and Pimental (1979) calculated that a 2.5 t/ha maize yield can be realised on 2 ha of land with an energy input of 2.2 M kcal/ha. To double the yield on the same land an energy input of 6.5 M kcal is required; to reduce land under cultivation by half and maintain yield, 3 times the energy input is needed.

Energy efficiency for irrigation varies widely among systems. Farm machines powered by fossil fuels have little effect on crop yields but reduce time spent in land preparation, planting and harvesting, helpful to farmers in ecologies with short growing seasons (Pimental 1984).

Energy in Post-Production Systems

Parikh (1986) estimates the energy consumed in processing of various commodities, and presents formulae by which to determine energy demand in different transportation systems. Parikh's observations relate to conditions in India, Pakistan, Burma and Sri Lanka. He concludes that energy in south Asian post-harvest food systems consumes 4 times the energy taken up in agricultural production. In the countries considered commercial energy provided between 60 and 90% of the energy taken up in processing and transport. Energy for transport varies according to the distances covered, the proportion of population in urban locations, and the fuel efficiency of transport vehicles.

Parikh indicates, in the countries studied, that domestic cooking consumes by far the highest proportion of the energy expended in the post-harvest systems. He concludes that all developing countries need to examine how efficiently energy is used in post-harvest systems. His data indicates how domestic energy use rises with income.

Needs and opportunities to use electrical power more conservatively and efficiently are evident in all regions. It is estimated in India that losses in transmission and by misuse account for 22% of all the power generated (Economist 19 March 1994).

Food processors and distributors are the largest single industry in the USA but consume only 2% of the total electrical power generated. The food industries consumption is equivalent to 1.5 quadrillion BTU or 1.4 Exajoules per annum. Put into a different context: per dollar of capital and salaries, an aluminum smelter spends $1.20 on energy; on a comparable basis, a frozen food processor pays barely 5¢ (Hulse 1993).

In all developing regions fast rising demands upon limited energy sources will conduce to competition among all sectors. The capacity to provide energy will persist as a primary constraint to development (Economist 1994).

BIOLOGICAL RESOURCES AND GENETIC DIVERSITY

The human race has progressed through three technological ages, sometimes coined as "revolutions": Mechanical, Chemical and Biological. First came steam, then internal combustion engines and electric motors to replace human labour and animal power. The Chemical age began in the 19th century with the German dyestuffs industry. The Biological gained momentum after the structure and means of modifying and transplanting DNA were discovered.

At the UN Conference on Environment and Development, most governments signed two Conventions: one on Climate Change, a second on Biological Diversity, the purpose of the latter being to ensure ecologically sound conservation of, and freedom of access to all natural organisms. Concern has been expressed at the economic and ecological consequences of the erosion of biological diversity by degradation of natural habitats, and by plant breeding and seed technology practices that may restrict conservation of and accessibility to germ plasm resources. There exist vast unexamined reservoirs of flora and fauna that could contribute to agricultural and industrial diversification. Total vertebrates and invertebrates are estimated at over 30 million species. Out of more than 250,000 identified plant species, 80,000 are believed to be edible, 150 are cultivated commercially, and three provide 60% of the world's human food energy. The 64,000 identified fungal species are less than 10%, and the 80,000 known arthropods barely 0.1% of all that are believed to exist (CABI 1991; Hulse 1993).

To collect, classify and conserve all living organisms is neither feasible nor practical, and to decide which species should be given priority is contentious between those whose interests are economic or aesthetic. It might be presumed that harmful pests and pathogens would be of lower priority than organisms that serve useful purposes.

Transgenesis

Some scientists argue for the selective preservation of only useful genes rather than whole organisms, the indicator of diversity being the evolutionary distance between related species. From this perspective, biologically beneficial genes are as useful from one organism as from several near-relations. Some plant breeders argue that science is only beginning its discovery of useful genes in wild species, land races and other organisms, and the transgenic means by which to transfer them into cultivated crops. Transgenesis is progressing in many directions and the future sources of commercially and socially useful genes cannot be foreseen.

Plant Breeding

Crop improvement, by plant breeding, depends on systematic utilisation of genetic diversity. Plant breeding is based on two basic operations: Selection and Generation of Variability. Selection identifies a plant or plants that possess genetically controlled desirable characters, characters that a plant can transmit to its progeny. Generation of variability comes from hybridisation: parents, which separately have several desired traits, are hybridised to generate a new population of progeny in which a spectrum of characters is embodied. Following hybridisation comes selection from among the new population of plants, those which display desired character combinations. These are then inbred followed by further selection until the ultimate character combination hoped for has been fixed and can be transmitted to future generations. Classical plant breeding of a new cultivar takes many years and calls for exceptional patience, persistence and perception. To accelerate the process, germplasm is moved back and forth among locations and latitudes to gain two or more crops each year.

If the purpose is to cross a tall red pea with a short white pea to generate a short red pea, the process is relatively straightforward. In most cases, however, a multiplicity of characters is involved, breeders seeking to combine such characters as yield, height, straw strength with resistance to several diseases, and properties that are technologically and nutritionally important. The difficulties of combining all desired characters in a single genotype are complex and to overcome them may take many years of patient persistent effort. To produce at least one plant that carries a combination of 21 characters in which the hybridised parents differ, more than 4 million plants must be grown in the second generation (IDRC 1986).

The history, principles and practices of plant breeding have been recently reviewed by Stoskopf (1993). Over time, plant breeders have used the most readily accessible sources of genetic variation, sources that can be incorporated in the shortest time. Genetic recombinations are most easily achieved by crossing plants of the same biological species which are genetically compatible and produce fertile progeny. Limited genetic diversity within a species restricts possibilities for genetic improvement realisable by classical plant breeding methods. Cross-pollination between plants of different species or genera tend to produce inviable, sterile progeny. As the genetic variability within a species is employed to exhaustion, breeders must seek and incorporate inheritable characters from other species of the same or unrelated genera. Modern techniques of wide-cross and intergeneric hybridisation are described and discussed in many recent scientific publications (Stoskopf 1993; IRRI 1984, 1985; IDRC 1986; APDC 1991; MSSRF 1991).

Sexual and Asexual Propagation

Plants can be classified into two main groups: those that reproduce sexually from seed, those that propagate vegetatively from plant tissue. Sexual reproduction by flowering plants occurs by fusion of a male gamete carried in pollen with a female gamete contained in the ovule. The resulting fertile embryo develops into a seed which when mature and planted in a favourable environment generates a root and a shoot to become a viable, fertile plant which in due season flowers and fruits. Self-pollination occurs when an ovule is fertilised by pollen from the same plant. Cross-pollination results when pollen is carried between different plants by birds, bees or the tools of plant breeders. Vegetative reproduction is an asexual process and can originate from bulbs, corms (swollen underground stems), rhizomes (stems that creep underground), tubers or stem cuttings. Any piece of plant tissue that is excised and used to regenerate a new plant is called an explant, and may include sizable cuttings, microscopic fragments of tissue, or isolated cells.

Tissue and Cell Culture

Recent years have seen considerable progress in the propagation of plants from somatic cells isolated from vegetative tissue. The explant is inoculated into a nutrient medium, a liquid broth or solid gel, containing nutrients and substances that stimulate cell division and proliferation. Dependent on the source of the cells, composition of the medium, ambient illumination and temperature, an explant may develop into a mass of undifferentiated cells - a callus, or into an organised pattern of differentiated cells which so arrange themselves as to develop a shoot and root. In some instances, by changing the composition of the nutrient medium, cells in an undifferentiated callus man be induced to reorganise themselves to develop into a plantlet with root and shoot.

The advantages of tissue culture include (a) production of a large number of plants from a single superior plant more rapidly than can come by seed multiplication; (b) generation of disease-free progeny from an infected parent; (c) reproduction from plants difficult to propagate sexually; (d) propagation from wide intergeneric crosses. Tissue culture can start with explants from virtually any part of the parent: from leaves, stems, roots, flowers, pollen, anthers, immature seeds, cotyledons or buds. There remains a good deal of empiricism in tissue culture techniques, the reasons why some species and media conduce more to propagation by tissue culture than others are not fully understood.

Individual plant cells can be induced to propagate by dispersion in a congenial medium. In addition to culture of intact cells, propagation of protoplasts, cells from which cell walls have been removed mechanically or biochemically, has progressed rapidly in recent years. Protoplasts can be modified in ways that are impossible with intact cells. Protoplasts can take up foreign genes by micro-injection or from organic vectors. Protoplasts from different plants can be fused to produce hybrids unattainable by sexual crossing.

Embryos generated by fusion of male and female gametes from widely different parents often are infertile and incapable of generating offspring, sometimes through incompatibility between the embryo and the endosperm. To avoid abortion and loss, the young embryo can be excised and grown in a culture medium. Embryo rescue techniques, known in the laboratory for many years, are now serving an important function in the synthesis of viable interspecific and intergeneric hybrids. Diverse physical and biological techniques now permit transfer of inheritable characters between genetically unrelated organisms to yield progeny with unique biochemical and physiological characters.

Transgenic Evolution

Deliberate modification of the genetic and biochemical constitution of plants and other organisms may be unwelcome to self-appointed guardians of "natural" organisms. It should be noted, however, that transgenesis as induced in the laboratory simply achieves rapidly what has come about by natural evolution over many millennia, as genes have jumped from one organism to another. Genetic mapping has revealed how bacteria have taken genes from and given genes to other microorganisms, plants and animals. A gene encoding for the enzyme glyceraldehyde dehydrogenase in certain *E. coli* strains is closely similar to a comparable gene in the human genome; a bacterial gene for glucose phosphate isomerase is identical to a gene found in certain food crops. Genome mapping is finding genes that have invaded and established themselves in foreign chromosomes, probably as with other biological evolutions to improve their chances of survival (Hulse 1993).

Mutations, sudden and inheritable modifications that occur in genes and chromosomes, can come about naturally or be induced by chemical or physical mutagens. Natural spontaneous mutations, followed by natural selection, have provided immense sources of genetic variability from which living organisms are derived. It was by spontaneous mutation over many millions of years that genetic diversification originated. Only in recent times have mutations caused by natural forces been enlarged and extended by controlled scientific interventions.

Melville (1966) suggests that the angiosperms, flowering plants whose seeds are enclosed within an ovary, originated from a common ancestor, the Glossopteridae that dominated the flora of Gondwanaland until that ancient land mass broke up and by tectonic drift formed the present continental system. Over many millennia these early progenitors mutated, hybridised and procreated into the countless crop genotypes which provide genetic diversity to modern plant breeders.

Origins of Diversity: Germplasm and Gene Banks

To ensure propagation and continuity of their species, wild plants scatter their seeds, a habit inconvenient in cultivated crops. An early objective of plant breeding was to propagate plants that held their seeds intact until harvested and threshed. Primitive plant species were carried across and between continents by birds, other organisms and by migrating humans. With each translocation, the species if it survived, would mutate and evolve to adapt to unfamiliar agro-ecological conditions.

Systematic collection of germ plasm, the range of genetic material essential to plant breeding, began in the mid-19th century at the Royal Botanic Gardens in Britain. About 40 years ago a Russian scientist, Vavilov, identified certain areas where exceptionally high levels of particular plant genetic variability were evident. These centres of genetic diversity are known as Vavilov Centres. For political reasons, Vavilov was discredited, but in 1991 a movement started to establish Vavilov Centres of Biological Diversity, the first being inaugurated close to Madras. Stoskopf(1993) illustrates nine recognised regions of diversity for 36 principal and industrial crops.

Plant genetic diversity is systematically protected and expanded in germ plasm and gene banks. Genetic conservation is broadly divided between *in situ* and *ex situ*, the two being illustrated in Table 35. *In situ* banks include natural wild habitats and farmers' fields. *Ex situ* banks embody botanical gardens; herbaria; field banks where materials are grown in outside plots; atmosphere-controlled seed banks; together with *in vitro* storage of vegetative tissue, pollen and excised DNA stored in bacterial or viral cultures. Cryopreservation is used to hold vegetative materials for long periods. Strictly speaking, a germplasm bank holds materials from which breeders generate progeny; a gene bank is where pieces of cloned DNA are stored (Plucknett et al. 1987).

Table 35. Biological diversity: germ plasm banks.

In situ	*Ex situ*
Natural habitats On farms (Traditional guardians were farmers)	Botanical gardens Herbaria Field gene banks Seed gene banks (Controlled atmosphere) *In vitro* storage Pollen storage DNA storage

The International Plant Genetic Resources Institute (IPGRI)

Germ plasm banks from which plant breeders draw their materials are held at international, regional and national centres and by private seed companies. The most important international institution, a member of the Consultative Group on International Agricultural Research, is the International Plant Genetic Resources Institute (IPGRI) formerly known as the International Board for Plant Genetic Resources (IBPGR) with headquarters in Rome (IPGRI 1993; IBPGR 1991).

IPGRI exists to advance the conservation and use of plant genetic resources for the benefit of present and future generations, particular concern being for developing countries' needs. IPGRI works as a partner with other international and national organisations, in research into the principles and practices of genetic resource conservation, and in training of people so engaged. IPGRI defines biological diversity as the total variability within all living organisms and the ecological complexes they inhabit. Added to its primary purpose of systematic expansion and conservation of plant germ plasm, IPGRI seeks to inhibit genetic erosion, the retrogressive loss that occurs in part by natural forces, more seriously by human predations.

IPGRIs principal partners are the 12 International Agricultural Research Centres (IARCs) of the CGIAR that are responsible for genetic improvement of food crops, forages and forest species. Table 36 lists the approximate numbers of accessions of major food crops held in the germ plasm banks of the CGIAR IARCs, the total being over 600,000.

Maintenance of a comprehensive germ plasm bank calls for exceptional skill and experience in modern taxonomic methods, new materials being acquired from collecting expeditions to sites of genetic origin. In its time IBPGR conducted over 300 expeditions to more than 90 countries in order to expand and diversify its international collection of valuable crop species.

Table 36. CGIAR collections.

Centre Crop	Accessions ('000)	Centre Crop	Accessions ('000)
CIAT, Colombia		ICRISAT, India	
Phaseolus beans	27	Sorghum	33
Cassava (*in vitro*)	5	Pearl millet	16
Forage crops	20	Pigeon pea	12
		Groundnut	13
CIMMYT, Mexico		Finger millet	3
Bread wheat	53		
Durum weat	13	IITA, Nigeria	
Triticale	13	Cowpea	15
Barley	8	Rice	12
		Yam	2
CIP, Peru			
Potato	4	ILCA, Ethiopia	
Sweet potato	5	Forage crops	10
ICARDA, Syria		INIBAP, France	
Barley	20	Bananas, plantains	0.5
Wheat	24		
Chickpea	8	IRRI, Philippines	
Lentil	7	Rice	81
Faba bean	4		
Forages	20	WARDA, Ivory Coast	
		Rice	6

The Rockefeller Rice Biotechnology Program

Modern methods of transgenesis - the introduction of alien genes into the nuclear DNA of established genotypes - has opened remarkable opportunities for plant improvement and genetic diversification. During the past 10 years, in cooperation with the International Rice Research Institute, the Rockefeller Foundation has invested substantially in an international research network by which to introduce useful foreign genes into the nuclear genome of rice plants, to ensure that the transplanted genes are inherited by sexual progeny in a conventional Mendelian manner.

Three procedures have been used to insert alien genes (1) DNA uptake by protoplasts is mediated through polyethylene glycol; (2) DNA is introduced to protoplasts by electroporation; (3) a particle gun is used to deliver DNA coated microprojectiles into the cells of immature embryos. Following insertion of alien genes methods are developed to regenerate transformed cells to produce transgenic rice plants, and to ensure each alien gene is integrated, transcribed and translated to synthesise the functional proteins desired. The ultimate purpose is to devise systematically a set of procedures to be applied routinely to generate transgenic plants covering a wide range of genotypes.

A specific objective is to transfer genes that encode for specific resistances to pests and pathogens from wild plants into cultivated species, to lower dependence on chemical pesticides. Aspects that relate to safety and nutritional quality are discussed later. In addition, the program is mapping the rice genome and studying the molecular biology of viruses, fungal and bacterial pathogens that infect rice plants. Techniques developed and knowledge gained in this innovative program will, it is expected, be adapted by others who seek useful transgenic derivatives from other crops of economic and nutritional importance.

GLOBAL FOOD RESOURCES

One of the most useful services provided by an international agency is the annual publication of crop, livestock and fisheries production estimates by the Food and Agriculture Organisation (FAO) of the United Nations. The statistics contained are best estimates since no agency can determine accurately the size of harvest from every farm, how many animals are owned and slaughtered in every rural community, and the weight of fish landed from net and line. Nor can there be a precise assessment of the magnitude of the physical, nutritional and economic losses that occur between the places of harvest and consumption.

Food Crops

Table 37 quotes recent estimates of global production of the main food crops. Cereal crops are raised on 48% of all cultivated crop land, wheat, rice and maize make up over 90% of the total cereals harvested. Table 38 gives estimates of cereal production in developing regions and the levels forecast for early in the next century. While opportunities to raise productivity of maize and other cereals may be optimistically viewed, there is evidence that the production increase of wheat and paddy rice may be less than came about over the past 30 years, from the time the short-straw wheat and rice varieties were developed and adopted. Among those listed as "Other Cereals" barley and sorghum are the most important in terms of scale of production, though, globally, relatively small proportions are consumed as food grains; sizable quantities of both barley and sorghum are converted to alcoholic drinks. In the USA, sorghum is grown mainly for livestock feed.

Amino acids in legumes and cereals are nutritionally complementary; about two parts by weight of cereal with one part legume provides a balanced protein. Legume production in Third World countries has fallen from 12 to 7.5 kg per capita per annum during the past 25 years.

As shown in Table 37 soybeans and groundnuts, grown mainly for oil, represent *ca.* 85% of total legume production. After extraction, residual meal is mostly fed to livestock. In India, home to many vegetarians, the ratio of cereal to legume production is of the order of 10 to 1.

Table 37. Estimated global food production 1990.

Cereals	Mt	Legumes	Mt	Root Crops	Mt
Wheat	595	Soybeans	107	Potatoes	270
Rice	520	Groundnuts	23	Cassava	170
Maize	475	Other legumes	25	Sweet potato	130
Total all cereals	1700	Total legumes	155		

SOURCE: FAO (1993).

Table 38. Cereal production: developing countries.

	Production (Mt)			Growth (%/yr)	
	1969-71	1988-90	2010	1970-90	1990-2025
Wheat	67	132	205	3.8	2.1
Paddy rice	177	303	459	3.0	2.0
Maize	70	112	196	2.7	2.7
Other cereals	68	85	134	1.3	2.2
Total	322	531	842	2.8	2.2

SOURCE: FAO (1993).

Among developing countries rice is the main cereal. Being an aquatic plant it is well adapted to the humid, high rainfall heavily populated ecologies of Asia. Wheat, maize, barley, soybean and groundnut production have been stimulated by research that has demonstrated how their vast genetic diversity and functional properties are adaptable to produce an impressive range of food products. No comparable research effort to expand their utility has been invested in sorghum or the thousands of legume genotypes held in international germ plasm collections.

Potatoes, for many years regarded as mainly carbohydrate calories, are now recognised as a source of good quality protein and several essential micro-nutrients. Progress in breeding of cultivars adaptable to tropical climates, and in propagation from true seed, as distinct from perishable seed tubers, have stimulated the spread of potatoes among developing countries where they find ready acceptance in fast food outlets that are proliferating rapidly in urban centres.

Cassava, the staple for over 200 million people, contains about 25% starch and 5% other carbohydrate and cellulosic matter. For poor people it serves as a reserve food crop that can be left in the ground until needed. In Asia and Latin America it is also grown as a plantation crop for industrial conversion to starch. The composition of starch with various reactive groups in the molecule is congenial to transformation to a diversity of chemical derivatives many used in food products, others in textiles, paper sizing and mineral extraction.

Livestock Products

Table 39 indicates how meat and milk production grew around the globe over the past 30 years. Disparate milk yields per animal among regions illustrate the opportunities for higher productivity through selective breeding and improved feed utilisation. Table 39A estimates expected growths in population and average income and their probable influence on demand for livestock products. Table 40 indicates the exceptional anticipated growth in demand for animal products in all developing regions, particularly in E and SE Asia. In light of the scarcity of arable land it is difficult to imagine how Asian nations can realise self-sufficiency in livestock production in the near term. Table 41 sets down demand for animal feeds among developing regions.

Table 39. Average annual meat and milk production (Mt/yr).

AVERAGE MEAT PRODUCTION	1960's	1970's	1980's
Total			
World	82	120	155
LDCs	28	41	56
Industrial nations	54	79	99
Beef			
World	32	45	50
LDCs	11	14	15
Industrial nations	21	31	35
Pork, ham, etc.			
World	31	45	62
LDCs	10	16	25
Industrial nations	21	29	37
Poultry			
World	12	23	35
LDCs	4	7	12
Industrial nations	8	16	23

AVERAGE MILK PRODUCTION	1960's	1970's	1980's
World	360	430	460
LDCs	69	94	80
Industrial nations	290	340	380

Average milk yield 1987 (kg/cow/yr)

World 2100	USA 6200	Canada 4600	EC 4000	USSR 2100
LDCs 740	Africa 340	Asia 700	L America 2200	

Table 39a. Developing nations: population and income.

	Population				Average Income/capita/yr	
	Total (Millions)		Urban (Percent)		$USD	Increase/yr (%)
	1990	2025	1990	2025	1990	1968-90
SS Africa	500	1300	29	54	450	0.2
ME/N Africa	250	500	57	75	1100	3.4
SE Asia	1700	2600	25	49	700	6.1
S Asia	1100	1900	30	56	300	2.0
L America	450	700	71	84	1800	1.9
Total	4000	7000	40	>60		

Forecast increase in meat/milk production 1990-2025 (%/yr).

	Red Meat	White Meat	Milk	Population increase
SS Africa	2.0	5.4	2.8	3.1
Asia	2.8	3.4	3.3	1.9*
L. America	0.6	2.5	1.7	1.8

* 1990-2025 Population of SE Asia forecast to increase by 70%; urban population to triple; rural increase 17%. Chinese population to increase 41%; urban population to double; rural 36% total.

Prospects for livestock production in developing regions has been sensitively reviewed in a Winrock International Foundation publication that compares differences in resources and opportunities among regions and estimates future demands up to 2025. The report reveals how rising demand is driven by three factors: population increase; demographic changes, mainly rural-to-urban migration; rising disposable incomes. It identifies as major constraints: insufficient good quality feed; low levels of animal health; too little development of superior animal breeds; inferior standards of livestock management. It predicts that by 2025 the Third World animal feed demand for metabolisable energy sources will rise by 77%, for crude protein by 74%, while demand for both will more than double in the Asian region (De Boer et al. 1994).

Table 40. Estimated livestock production in developing regions (Mt/yr).

	Beef	Sheep/Goat	Poultry	Pig	Total meat	Total milk
SS Africa						
1987	2.1	0.9	0.9	0.4	4.3	9.6
2025	4.5	2.0	7.8	1.2	15.5	22.9
M East/ N Africa						
1987	1.1	1.4	1.4	-	4.1	12.1
2025	2.4	3.4	10.0	-	16.3	27.2
Asia						
1987	2.1	2.0	4.7	21.0	30.8	68.2
2025	8.4	4.3	28.6	64.6	107.5	217.5
L America						
1987	8.5	0.4	4.4	3.0	16.29	39.9
2025	10.5	0.7	13.8	5.5	30.5	73.8
Total						
1987	13.8	4.7	11.4	24.4	55.4	129.8
2025	25.7	10.5	60.3	71.3	169.8	346.4

N.B. Total meat includes other animals not listed. Total milk includes cows, goats, sheep, buffalo.

SOURCE: FAO (1993).

Table 41. Animal feeds demand (Mt).

	1970	1980	1990	2010
All LDCs	56	114	160	340
N East/N Africa	10	19	32	65
E Asia	21	49	73	174
China	16	39	53	129
L America/ Caribbean	22	38	45	79

Fisheries

Fisheries provide food and employment for many Third World people. It is estimated that of roughly 13 million fisherfolk across the world, over 11 million live in developing countries.

Table 42 indicates how global fisheries production rose rapidly between 1950 and 1970, but now has reached a plateau of barely 100 Mt which several stock assessment authorities consider as the maximum sustainable harvest from world oceans. Table 43 quantifies declining fish stocks in the Atlantic Grand Banks, for many centuries a prolific fishing ground for European and North American nations. Table 44, entitled "Overfishing", sets out the main causes of aquatic resource depletion.

Table 42. Annual global fisheries production.

	Total (Mt)	Non-food (Mt)	Per capita food (kg/yr)
1950	20	3	6.7
1960	39	11	9.0
1970	65	25	10.9
1980	72	21	11.3
1989	100	30	13.2
1990	97	31	12.3
PERCENT CHANGE			
1950-60	95	270	36
1960-70	66	124	20
1970-80	11	(16)	4
1980-90	35	43	8
1989-90	(3)	3	(8)

Production = Landed live-weight equivalent.
Per capita food = (Total production - Non-food) ÷ Population.
In Canada (1988) 21 kg/capita food = approx 6.8 kg/capita edible mass.

SOURCE: FAO Production Year Books of Fisheries 1989, 1990, 1991, 1992.

Of the total global catch, 88% is from capture fisheries and 12% from aquaculture. Of the capture harvest 80% comes from inshore, 12% from inland fresh waters and *ca.* 8% from the deep oceans. Artisanal fisheries with over 90% of fisherfolk produce less than 50% of the total catch. FAO predicts that to meet future demand production must almost double to 200 Mt by the year 2025. Since the harvest from the oceans may have reached its limit, best hope lies in enhanced fisheries that involves captive breeding and mass propagation in natural environments and fish ranching; and aquaculture - the systematic farming of various aquatic organisms. Highest per capita production, by capture and aquaculture, is among Asian nations.

Table 45 indicates the state of aquaculture. The Peoples' Republic of China, where fish farming has been practised for many centuries, remains the biggest producer, though all other Asian nations are expanding and diversifying production. Salmon farming has grown rapidly in Europe and North America in recent years. Gravid females can be induced to oviposit by gonadotropin treatment; growth rates and body size can be modified by biosynthetic hormones and by sex inversion (Hulse 1993).

Fish for food could be increased by reducing losses after harvest, estimated at 4 Mt per year, landing the by-catch discarded by shrimp trawlers and the Asian deep sea trawlers that use 50 km long drift nets, and by upgrading the pelagics now converted mainly to oil and fish meal.

Table 43. Fish stocks - Canadian Atlantic: Labrador, Newfoundland, Grand Banks.

Cod	1968	1980	1991	Mean (1960-91)
Catch ('000 t)	810	200	120	363
Biomass ('000 t)				
3-yr-old	2500	750	1000	
7-yr-old	1600	250	100	
Haddock	1960	1985	1992	
Catch ('000 t)	80	10	1	

Table 44. Overfishing and unsustainable waste.

CAUSES:	Too many people fishing: uncontrolled licensing and access Wasteful technologies (driftnets); discard of by-catch Inadequate stock assessment and aquatic resource management Excessive post-harvest loss of fresh, smoke and salt-cured
BIOLOGICAL LOSS:	Depletion of species, juveniles, broodstock
ECONOMIC LOSS:	Cost of fishing effort exceeds economic yield.

Table 45.

	Aquaculture 1990	
	Production (Mt)	Percent
World	12.0	100
Asia	9.5	80
Europe	1.5	12
North America	0.5	4
Africa	0.25	2
Latin America	0.25	2
CLASSES:	World (%)	Asia (% of World)
Finfish	50	61
Molluscs	16	14
Crustacea	6	5
Aquatic plants	28	20

1990:	Aquaculture provided 12% of global fish production
2000:	Aquaculture must provide at least 20% of production
1975-90:	Annual average increase *ca.* 5.5%/yr.

Comment on Fisheries

Recognising that harvest from capture fisheries have reached and probably exceeded their sustainable limits, greatest hope for future fisheries rests upon aquaculture in fresh waters, both natural and in constructed enclosures; and in mariculture in coastal and brackish waters.

In the Changi Strait of Singapore, scientists report annual harvests from mussels cultured under floating rafts that yield 90t of protein per hectare of surface water. Scottish salmon farms increased their production from 1100t in 1979 to 33,000t in 1989. Expansion of salmon production is reported from Canada and Norway, of shrimp from Ecuador and several Asian countries.

Donaldson (1986, 1988) and others describe how gravid female finfish are induced to oviposit by injections of modified and mixed forms of gonadotropin releasing hormones; how ovulation and spermiation may be stimulated by various extracted and modified hormones. Rates of growth in Pacific chinook, rainbow trout, catfish and other commercial species can be accelerated by avian and bovine growth hormones synthesised by genetically modified bacteria. Sex inversion of newly hatched chinook alevins by androgen treatment provides a cultured population that is predominantly female. Females of chinook and several other species are of higher commercial value than males because of more desirable body conformation and for the roe they provide. It is reported from Russia that caviar roe can be excised from sturgeon by Caesarian surgery, the post-operative females being then released to produce more roe.

Limnological studies suggest that many of the *ca.* 600,000 km^2 of constructed lakes and reservoirs in developing countries, if freed of pollution, are amenable to various systems of fish farming; the water subsequently being continually passed through turbines to generate electricity and/or conveyed elsewhere to irrigate farm land.

Though it is unlikely that aquaculture and mariculture will in the foreseeable future entirely compensate for the losses from capture fisheries, if adequately supported by research and efficient post-harvest means of preservation and distribution, fish-farming in clean, unpolluted coastal and inland waters can contribute substantially to commercial aquatic resources in many regions.

CHAPTER 5. AGRICULTURAL RESEARCH

Agricultural Research in the USA

In 1982, the Rockefeller Foundation and the Office of Science and Technology Policy held a workshop to discuss the critical issues that affect American agricultural research and concluded the following: Related to public policy: agriculture is vital to the U.S. economy in that agricultural commodities account for close to 20% of all exports; the food and agricultural sector is the largest of all U.S. industries and employs 20% of the civilian labour force. Annual rates of return on research expenditures in agriculture range from 35 to 50%, higher than returns to most other public sector investments (Rockefeller 1982).

Constraints were given as: (a) No coherent national agricultural policy; (b) Lack of clear implementable national goals, priorities and plans for agricultural research; (c) Little public understanding of the importance of agriculture to U.S. strength and vitality and the well-being of American citizens; (d) A changing political balance of power in Congress from primarily rural to predominantly urban, which portends even greater difficulty in rationalising public policies related to agriculture and agricultural research.

While at least 10 federal agencies, in addition to USDA, invest in research related to agriculture, financial constraints and inadequate support have encouraged parochialism, resistance to change and self-protection among academic and other research institutions. A low level of importance is evident in the relatively few agricultural scientists in the National Academy of Sciences. It is unclear who can speak with authority for agricultural research and the agricultural system.

In relation to research performance, the report noted an unevenness of focus and a piecemeal approach to basic research for fundamental knowledge on which the future of agriculture will depend. It noted deficiencies in the quantity and quality of basic research traceable to (1) lack of a clear priority for basic research; (2) increasing substitution of political for scientific goals by the Congress; (3) insufficient public financial support for agricultural research.

The report recommends: strong sustained federal support of research to develop new knowledge and new technologies, with vigorous support at both state and national levels. Long-term agricultural issues need to be analysed and related to a broader social context. Finally, there is need to stimulate and coordinate interactions between American and international agricultural scientists (Rockefeller 1982).

Relevant to the last, the U.S. government seems disposed to reduce support for two international cooperative programs to which several U.S. universities have impressively contributed over many years.

The USAID Cooperative Research Support Programs (CRSP) finances consortia of U.S. universities to work cooperatively with agricultural scientists in Third World institutions. Typical is the INTSORMIL CRSP in which five U.S. universities cooperate with scientists in Africa and Latin America to improve and expand production and utilisation of sorghum and millet, both are important feed grains in the USA, and vital subsistence cereals in semi-arid Africa. The results therefore benefit both the U.S. and the poorest Third World people (CRSP/USAID 1993).

The Consultative Group on International Agricultural Research (CGIAR) and its family of research centres appears also under threat of reduced U.S. financial support. The CGIAR is discussed in greater detail later.

Lacy and Busch (1984) also comment critically on government policy related to agricultural research in the USA stating that little yield increase has been seen recently in most major U.S. food crops. Other proclivities antagonistic to agriculture include conversion of over 400,000 ha of prime cropland every year to non-farm use; cultivation of marginal lands and fragile soils with higher use of fertilizer and irrigation water; over-drafting of aquifers in several states; transformation of many small farms into large specialised, capital intensive production units. Since the early 1930's, the number of farms fell from 7 million to 2.3 million. Over 50% of total harvested value is now produced by 7% of the farms.

Lacy and Busch (1984) list as essential to national food security: technologies conservative of energy and non-renewable resources; that sustain food and fibre production; are non-polluting and not confined to large-scale production; that address social issues of equity, quality of life and human development. All wait in need of more research.

The USA cannot for ever be the bread basket of last resort for all the world; the demand on essential resources would be unsustainable. It is in the interest of the U.S., Canada and others who produce commodities surplus to their needs, to assist poorer nations to improve their food and agricultural productivities.

Traditionally, universities have pursued curiosity-inspired research and teaching, government institutions whatever bears upon the national economy, health and welfare, with the private sector concentrating on commercial products and processes. Universities who seek financial support from the private sector may neglect fundamental research. Lacy and Busch (1984) quote U.S. Congressman George Brown: "Agricultural research is going through a mid-life crisis. We've emphasised applying existing knowledge and failed to replenish our intellectual capital".

Lacy and Busch (1984) cite evidence to indicate that agricultural scientists are an aging community. In a 1978 Report of Characteristics of Persons in Fields of Science and Engineering, the U.S. Bureau of the Census of the Dept. of Commerce, stated that among all scientists and engineers, agricultural scientists were of the highest median age group (45.3 years). The U.S. Office of Technology Assessment in 1981 reported that barely 2% of scientists in USDA were aged under 30 years, while 39% were over 50 years. It is suggested that as scientists age, they become less imaginative and innovative, more conservative and less willing to take risks; inexpert in information, data handling and other modern techniques.

Lacy and Busch express concern at the fragmentation of agricultural research: the tendency for each specialised discipline to operate in isolation is not conducive to sustainable holistic systems needed to maintain food security. Problems, opportunities and constraints are more easily defined in unidisciplinary than multidisciplinary parameters. Research on a single crop being more easily conceived and conducted than studies of mixed cropping systems, intensive monocropping is most rewarding of research effort.

While much lip service is paid to "alternative production systems" that are more conservative of natural resources, less dependent on agricultural chemicals, irrigation and capital investment, political dispositions and institutional organisations favour continuation of intensive agriculture, with more government investment in subsidies and price supports than in the long-term research needed to introduce the technological and agro-economic changes that many observers advocate (Lacy and Busch 1984).

International Agricultural Research

In 1989 the Technical Advisory Committee of the CGIAR published its perceptions of the needs and priorities for international agricultural research programs (CGIAR/TAC 1989). The TAC report expresses particular concern for production systems that are sustainable, and that ultimate products confer equitable benefits. It is remarked that in well-endowed regions the large and prosperous farmers are best equipped to adopt new technologies. Eventually the less affluent, smaller-scale producers adapt and adopt agricultural innovations that are neutral of scale. In the long run rate and extent of new technology adoption may be more influenced by ecological considerations than by farm size. The report comments on systems regarded by some as unsustainable, particularly those heavily dependent on agro-chemicals and mechanisation. It does not, however, equate what is unsustainable with intensive production systems, and suggests that despite mismanaged irrigation, and wasteful use of chemicals, when prudently conceived and controlled intensive systems that sustain high productivity are essential if food security is to be assured for rapidly expanding populations.

TAC recommended as priorities: research to raise the productivity of systems that can be sustained in adverse environments, systems that are conservative of natural resources; research to restore land that has suffered degradation; research on advanced techniques of molecular biology and genetic modification. The TAC report emphasises that donors who finance the research proposed must not expect rapid, spectacular results and benefits. Enhanced productivity that is ecologically and economically sustainable cannot be realised quickly. TAC underlines the urgent need for methods to measure the long-term value of research as it relates to productivity and natural resource conservation.

Subsequently TAC recommended that CGIAR support several eco-regional research programs, each dedicated to a particular ecology critical for agriculture. One such, the Desert Margins Initative, is referred to elsewhere in this publication (TAC/CGIAR 1990, 1994b).

In a report on food security to the United Nations Conference on Environment and Development, a Committee of the International Council of Scientific Unions sets out its perceived priorities for food and agricultural research (CASAFA 1991). The report deals with food security in broad perspective by addressing both production and post-production systems, the latter being integral of all that is involved between the products of farms and fisheries and the consumers they are to nourish. It deals with food quality and safety, the disposal and utilisation of waste and by-products of agricultural production and agro-industries. The report points to a pressing and persistent need to:

- Increase crop yields and farm productivity in developing regions;
- Protect and enhance the essential natural resource base: farm land, surface and ground water, genetic diversity;
- Reclaim and restore fertility in lands lost through misuse;
- Invest more in agricultural research and training, and in studies and development to integrate improved production with post-production systems and to stimulate rural agro-industrial employment;
- Devote more scientific resources to land capacity classification, to soil and water management research, and to determining land use that is economically productive and ecologically conservative.

The CASAFA paper points to several issues in need of consideration and clarification, and of substantial research investment:

- Contradictions abound among the many definitions and prescriptions for sustainable agriculture.
- It is unhelpful to propose any particular production system as globally sustainable. No system can satisfy all needs and ecologies.
- Scientists and donor agencies should not impose upon poor farmers in

developing nations, systems designed to remedy wasteful practices in more affluent agricultural economies.

- Though industrial nations could without undue deprivation reduce their reliance upon agricultural chemicals and other external inputs, developing nations' farmers, to keep pace with demand, need to increase crop production by at least 3% per year, a target unattainable from closed farm systems independent of external resources.
- Most proposals give greater emphasis to production technologies than to policy, social and economic constraints.
- Practically no attention is given to post-production systems: to the technologies, logistics and economics of preservation, processing, transport, distribution and utilization. Soil-to-mouth subsistence systems cannot satisfy the needs of urban and off-farm rural people who comprise over 40% of all developing nations' populations.
- Too little concern is shown for the diverse human resources and skills needed for research, training, design, development and management of ecologically protective production and post-production systems, and for the decrease in the proportion of populations who work on farms.
- Comparatively few who write and recommend lower-input, more labour-intensive production systems appear to be developing country farmers. Significant change can impose significant risk, risk that is borne by farmers. All agriculture is susceptible to risk. To reduce risk, changes to established production systems should be based upon sound research and thorough on-farm evaluation. Change must be introduced cautiously, with due concern for relevant economic, social and human factors.

Despite these and other related recommendations, support for research to raise agricultural productivity, to improve preservation, processing and distribution systems tends more to decline than increase.

Agricultural Research in Africa

Steadfast support for food and agricultural research is essential in every nation: to develop systems and technologies by which economically to satisfy expanding and changing demands; to induce resistances to persistently mutating pathogens and parasites, and to new pest races as they are carried from one continent to another. Food and agricultural research must be as persistently dynamic as the diverse biological, social and logistical constraints to be overcome and the opportunities that are to be realised.

As has been illustrated, the magnitude of need and demand for food in sub-Saharan Africa is racing faster than supply. More, not less, research being needed to reverse this pattern, a recent study of agricultural research in Africa makes for melancholy reading (Pardey et al. 1994). During the past 30 years, the number of professionally trained African agricultural scientists has risen 6-fold.In constant currency, investment in agricultural research, throughout the region, has fallen consistently over the past 15 years, the net result being scientists qualified to work effectively, but deprived of the resources to do so.

Pardey et al. (1994) report that between 1986 and 1991, investment in agricultural research by African governments, as a percentage of the total, dropped from 56 to 49%, while donor support rose from 35 to 45%. The net flow of Overseas Development Assistance (ODA) from donors to developing countries regressed from $61 billion in 1992 to barely $56B USD in 1993, about one-third going to Africa. Donor intransigence and insensitivity to the chronic and compounding difficulties of debt and deprivation augur dismally for future food security in Africa.

CONSULTATIVE GROUP ON INTERNATIONAL AGRICULTURAL RESEARCH (CGIAR)

A history of the foundation of the CGIAR was published by the World Bank (Baum 1986). It was inspired by agricultural research sponsored by the Rockefeller and Ford Foundations. In 1943 Rockefeller began a bean and cereal improvement program in cooperation with the government of Mexico, a program that generated semi-dwarf high yielding wheats, and eventually metamorphosed in 1963 to become the International Maize and Wheat Improvement Centre (CIMMYT). In 1960, the two foundations created the International Rice Research Institute (IRRI) in the Philippines; later in the 1960's they established the International Institute for Tropical Agriculture (IITA) in Nigeria, and the International Centre for Tropical Agriculture (CIAT) in Colombia.

These pioneering innovations made clear to the international donor and development community the opportunities and benefits to be gained from agricultural research designed to generate crop types and systems of production capable of raising yields and outputs from croplands in developing regions. A series of meetings in 1969 and 1970 led, in 1971, to a coming together at the World Bank of a consortium of donor and development agencies, an informal group, the CGIAR. To enable the CGIAR to function the World Bank provided a supporting secretariat. During its initial meetings the CGIAR created a Technical Advisory Committee for which FAO housed a secretariat in Rome. The CGIAR, created as an informal body without a legal constitution, gained credibility when the World Bank, the United Nations Development Programme and FAO became its joint sponsors. The CGIAR started by dedicating $12 million to support the four existing Centres: IRRI, CIMMYT, IITA and CIAT. During its first formal session in December 1971, the 11 donor members pledged $15 million. By 1976 CGIAR membership had risen to 26 donors who provided a budget of $63 million to the four original and seven new centres.

In 1992 the CGIAR, composed of 40 donor members who collectively provide close to $300 million, maintained a network of 16 international centres and related programs. The Group began as a small informal intimate body which in its formative years functioned with relatively little bureaucratic overload. Members dealt with the issues before them without formal speeches. CGIAR now convenes two meetings each year, the main gathering at the World Bank now resembles a United Nations agency General Assembly. The weight of documents is formidable, and much time is taken up with evaluations of Centres' activities. Nevertheless, each of the Centres fulfils an essential and vital research function, a function no other agencies are equipped to undertake so effectively. Detailed accounts of the activities of the CGIAR, its secretariats and its family of members are recorded in annual reports available from the CGIAR Secretariat in the World Bank (CGIAR 1972 to 1994).

At its beginning it was decided the CGIAR program would concentrate on research to improve food crops on which developing countries most depended. Research on rice, wheat and maize was already underway. Early in its history the CGIAR brought into its orbit the International Potato Centre in Peru. By collaboration between the Ford Foundation and the International Development Research Centre (IDRC) the CGIAR brought into being the Institute for Crops Research in the Semi-arid Tropics (ICRISAT) with headquarters in India and research responsibility for sorghum, millet, groundnuts, chickpea and pigeon pea.

The primary purpose of each Centre's crop improvement program is to maintain a comprehensive germ plasm bank that embraces a broad range of genetic diversity among the crop species within its mandate. Using classical and modern breeding techniques and collection expeditions the germ plasm banks are expanded and diversified (IFPRI/RF 1992).

Each of the crop improvement centres acts like a grocery store from which national research programs in developing countries can select genotypes and phenotypes of crop species best suited to their needs and ecologies. Each Centre conducts research better to understand and expand utility of its collections and to develop agronomic production practices best suited to the various agro-ecologies in which the crops may be cultivated. Each Centre offers training in the disciplines and activities that relate to crop improvement and utilisation. Since its inception, the CGIAR and its family of centres have assembled the world's largest and most diverse collection of plant genetic resources amounting in all to over 600,000 individual classified accessions. CGIAR's progress and priorities are periodically reviewed (TAC/CGIAR 1994a,b,c; CGIAR 1995).

International Rice Research Institute (IRRI)

IRRIs early history is recorded in a publication to commemorate its 20th anniversary (Chandler 1982). More than 90% of the world's rice is grown and consumed in Asia where demand is rising at over 2.5% a year. The fact that Asian rice production doubled during the 1960's and 1970's is largely attributable to IRRIs development of high yielding, short straw phenotypes. Now because both yield increases and the area planted to rice are tapering off, the population among Asian rice eaters is growing faster than is rice production. A cooperative research program between IRRI and the Rockefeller Foundation, which seeks to improve resistance to various pests and pathogens by transferring genetic resistances from wild species into cultivated rice has been described.

In addition to IRRIs many improvements in rice germplasm, several years ago IRRI scientists laid the foundation for modern concepts of farming systems research. The IRRI rice-based farming systems program began modestly with a few Filipino farmers, since when it has spread across over 12 Asian countries.

Contrary to the traditional process whereby agricultural research starts on a experimental farm, results from which are eventually transmitted to farmers through extension services, the IRRI program began by determining first the opportunities for and constraints to increased production on Asian farms. The methodology called for an analytical assessment of the technological, physical, ecological and economic resources among different farming communities, by what means could more productive technologies be introduced with least risk to the farmers. This methodological approach, which bears resemblance to marketing research techniques, has contributed extensively to increased and conservative production (IDRC 1981; Chandler 1982).

IRRIs present program is organised according to the ecosystems in which rice is produced, including irrigated, rain-fed lowland, upland and deep water cultivation. Each sub-program pursues germplasm improvement; increase and stabilisation of yields; enhanced resistance to pathogens, pests and adverse environments; resource conservation and ecological protection. As is common to other CGIAR centres IRRIs research program is cooperative with institutions in many rice growing countries.

For the future IRRI has identified critical research issues, first and foremost being the need to break through the present yield barrier by designing a new rice plant with a yield potential of 13-15 t/ha. The proposed characters include 3 or 4 large panicles each with 200 or more grains, no unproductive tillers, a sturdy stem, thick dark green leaves, and a vigorous root system. A concomitant approach is to develop vigorous hybrids borrowing from experience in China where, in 1992, over half the rice land was planted to hybrids that yielded 20% more grain than traditional in-bred types.

IRRI is also seeking improved nitrogen use efficiency, since 25% of increased production in the 1960's and 70's came from higher use of nitrogen fertilizers. Because the opportunities to expand land under rice are severely limited and heavy applications of agro-chemicals carry ecological and economic disadvantages, the need is urgent to raise the efficiency with which rice plants take up and use nitrogen. Other objectives include raising intrinsic resistance to pathogens and pests; determining the consequences of higher ultraviolet radiation and possible global warming to which rice may contribute by emissions of methane. Recent studies estimate methane emissions from rice fields up to 100 Mt per year, a level corresponding to 30% of total anthropogenic methane emissions.

Under the broad umbrella of "rice biotechnology", IRRIs research is progressing through embryo rescue in the transfer of genes from wild to cultivated species; anther culture to produce haploids and stabilise new lines; accelerating breeding programs by using genetic markers in place of phenotypic selection; DNA finger-printing to identify genetic variability among pests and pathogens; the introduction of novel genes through protoplast transformations. These are techniques now used by most CGIAR centres to improve the crops within their mandates.

Centro Internacional de Mejoramiento du Mais y Trigo (CIMMYT)

Maize and wheat, the crops of the International Maize and Wheat Improvement Centre, with its head office and research station in Mexico, together account for more than 40% of the cereals grown in the Third World. Some 50 Mha are sown to wheat derived from germ plasm improved at CIMMYT, over 8 Mha are under maize varieties developed from CIMMYT germplasm. CIMMYTs stated research objective is "To increase the productivity of resources committed to maize and wheat in developing countries, while protecting natural resources."

The yield gains in wheat achieved by reducing the height from close to 300 to below 200 cm, and shifting phytomass from the straw into the grains is well publicised. CIMMYTs semi-dwarf wheats form the basis of higher productivity in many countries. Other CIMMYT accomplishments less widely proclaimed are: (a) development of triticale, the first commercially successful intergeneric hybrid produced by plant breeding in which wheat and rye were hybridised to combine desirable characters from each species into a new genus; (b) stabilisation of a gene that encodes for higher lysine to enhance the biological value of maize protein (Hulse and Laing 1974; Cantrell 1990). Both seem to have achieved widest application in animal feed: triticale is a productive crop for forage and feed grain; high-lysine maize produces faster growth rates in hogs.

CIMMYTs primary objectives are to raise cereal yield potentials and to broaden their agro-ecological adaptation, the latter being served by CIMMYTs international testing nurseries where new lines are assessed for yield and stability in an extensive range of ecologies. CIMMYTs wheats are tested in over 30 international nurseries; in the late 1980's close to 800 experimental variety trials were carried out on maize.

As with all food crops the most pernicious inhibitors of higher productivity are hundreds of pathogens and predatory pests that cause substantial loss before and after harvest. Stem, leaf and stripe rust are the most destructive to wheat, so that a main germplasm improvement objective is to raise intrinsic resistance to rust races, a process that calls for dynamic vigilance in that fungal species can mutate to overcome inbred resistance faster than new resistance is introduced. While troublesome pathogens and pests can be carried from one region to another, particular pests tend to particular ecologies thus requiring that CIMMYTs improved germplasm be crossed with varieties established locally that have evolved resistance to local pests.

Drought stress is of unique severity in at least 40 countries where wheat and maize are cultivated. Breeding wheat for dry environments is undertaken cooperatively between CIMMYT and ICARDA in the low rainfall areas of the Middle East and North Africa.

As are IRRI and other crop improvement centres, CIMMYT is exploring modern transgenic techniques to synthesise wide crosses, to transfer resistances and other desirable characters from distantly related and wild species. CIMMYTs Molecular Genetics laboratory uses molecular markers to map the genomes of maize, wheat and other species used in hybridisations (CIMMYT 1992).

International Centre for Agricultural Research in Dry Areas (ICARDA)

ICARDA, established in 1975, with headquarters in Syria, carries out research in the Middle East and North Africa covering 22 nations and an area almost twice the size of the USA that is characterised by low rainfall, much being near-desert. The ICARDA region suffers from the world's most extreme food deficit, a population growth rate of 2.6% per year, and an expected 1.5 billion inhabitants by 2030. ICARDA bears the global responsibility for barley, lentils and faba beans, with regional responsibility for wheat, chickpea and several forage crops. By the end of 1989, over 150 improved crop types had been released, more than 3000 scientists and technicians from the region had attended ICARDAs formal training courses.

Centres for Tropical Agriculture (CIAT and IITA)

The International Centre for Tropical Agriculture (CIAT) and the International Institute for Tropical Agriculture (IITA) were initially intended to improve production in specific tropical ecologies: IITA, the low humid tropics of Africa;

CIAT, the infertile, acid soils of Latin America. While each centre has studied farming systems in its local ecology, most progress has been evident in particular crop improvement.

CIAT has improved the germplasm of Phaseolus beans and cassava, of pasture grasses and legumes for the acid-soil savannas. Bean cultivars, with greater pest and disease resistance and tolerance to drought, have been released to over 20 Latin American and African countries. CIATs cassava program was at first linked with swine production, but more recently has progressed in technological innovation for small-scale production, efficient processing and utilisation. CIATs tropical pastures program has assembled and classified an extensive germplasm bank of forage grasses and legumes, which in different environments have been tested in single and mixed swards to determine sustainable stocking densities for grazing livestock.

IITA collaborates with CIAT in cassava improvement for Africa, while in its own right studying yams and plantains. IITA is recognised for its biological control program which identifies, propagates, releases and monitors natural enemies of the cassava mealy-bug, a pest brought into Africa by carelessness, causing widespread crop devastation.

Equally notable are IITAs improvement of cowpea, the most important food legume of West Africa's low rainfall areas, and its demonstration of alley cropping, a form of agro-forestry wherein food crops are grown between rows of leguminous trees whose roots, in symbiosis with soil bacteria, convert atmospheric nitrogen to an assimilable nutrient.

Centro Internacional de la Papa (CIP)

The International Potato Centre, with its headquarters in Peru, began in the early 1970's by combining two potato research programs, supported respectively by Rockefeller Foundation and USAID. CIP joined the CGIAR soon thereafter. The potato was traditionally regarded as a temperate climate crop until CIP demonstrated the opportunities for production in tropical ecologies, where until CIPs alternatives, potatoes were propagated from seed tubers imported from Europe and North America.

CIP has improved storage techniques for seed potatoes and developed propagation from true seed. Potato cultivation consumes exceptional quantities of chemical pesticides. CIP is lowering the dependence by biological controls that use various funghi, viruses, insects and sex pheromones to attack pests troublesome to potatoes.

Recently CIP assumed global responsibility for sweet potatoes, a crop of economic and nutritional significance, being high in carbohydrate, protein and vitamin A. To facilitate global distribution of improved germplasm CIP is generating disease-free material by tissue culture.

International Network for the Improvement of Banana and Plantain (INIBAP)

INIBAPs foundations were laid by the International Development Research Centre (IDRC) which recognised bananas and plantains as food for many poor people, not simply as an economically important export crop. Total global production of bananas and plantains is estimated at 60 Mt per year; roughly 8 Mt of dessert bananas being exported. INIBAP seeks to stimulate and coordinate exchange of *Musa* germplasm and with in vitro collections, to maintain a *Musa* conservation network. *Musa* is the genus to which both bananas and plantains belong.

INIBAP functions as integrator of several networks and collaborating institutes with a common major objective of identifying and inducing resistance to Black Sigatoka, a disease that devastates *Musa* species in Asia, Central and South America. INIBAPs regional networks are located in Latin America, the Caribbean, West and Central Africa, East Africa and Asia, each acting to expand and maintain a germplasm collection of *Musa* species of economic importance.

West African Rice Development Association (WARDA)

WARDA is somewhat unique in being originally created by governments of 17 African rice growing nations, where the demand for rice has risen at 4% per year over the past two decades. It has been demonstrated that for economic and ecological reasons, production technologies developed on other continents are not readily transplanted into Africa, and that research is needed to develop rice cultivars and cultivation systems acceptable to African producers. WARDA scientists work to improve rice germplasm, production systems under irrigation, in upland areas and in mangrove swamps.

International Livestock Research Institute (ILRI)
International Livestock Centre for Africa (ILCA)
International Laboratory for Research on Animal Diseases (ILRAD)

In the early 1970's CGIAR created two centres for research on African livestock. Two parasitic diseases: trypanasomiasis and theileriosis (East Coast Fever) afflict cattle in one-third of Africa. Both are carried by arthropod vectors: trypanasomiasis, endemic in 37 countries, is carried by the tsetse fly, East Coast fever, serious in 11 countries, is a tick-borne infection. ILRAD was created to devise effective and economic methods of alleviation and has concentrated on immunological research, on diagnostics and chemotherapeutic methods of control.

At its main station in Nairobi, ILRAD has made immense progress in gaining a comprehensive understanding of bovine immune systems and in identification and transfer of genetically controlled resistances to these devastating diseases.

ILCA based in Ethiopia has critically examined the African ecologies where livestock are raised, assessed and analysed the social, economic and technical constraints to and opportunities for greater meat and milk output from cattle, goats and sheep. Collection and improvement of forage crop germplasm and the utilisation of such crop by-products as maize, sorghum and millet stover are facets of ILCAs program. Genetic control of the lignin content in sorghum stalks, makes possible the use of the upper stalk, low in lignin, for livestock feed; the lower stalk, higher in lignin is better used as mulch to improve water retention in the poor soils of semi-arid Africa.

In 1994, ILCA and ILRAD were administratively fused into a new International Livestock Research Institute (ILRI) to undertake research essential to all developing regions.

International Centre for Living Aquatic Resources Management (ICLARM)

ICLARM is a recent addition to the CGIAR family. It was established in the Philippines 15 years ago to carry out research on capture and culture fisheries, and the management of coastal ecologies and aquatic resources. Aquatic resource management suffers several disadvantages relative to agriculture. Crop cultivation and livestock husbandry have been practised for about 10,000 years. Until recently, fish culture was confined to a few countries where the production sites were stocked by naturally occurring juveniles. Numbers of food fish species are greater than husbanded livestock by many orders of magnitutude; the ecologies in which they exist are more diverse and less well understood than those where crops are cultivated and livestock raised.

ICLARMs program includes resource conservation: protection of coastal and inland aquatic systems; ecological characterisation in relation to fish farming and alternative water uses; conservation of habitats and species calling for identification and protection of biodiversity; conservation of natural fisheries environments; assessment of natural fish populations and their depletion by capture fisheries.

ICLARM studies coastal systems and how they and the resident natural organisms are affected by human activities, agricultural and industrial contaminants, and domestic settlements. By the year 2010, three-quarters of the world's population will live within 40 km of an ocean coastline. The effects of the pollution on coastal waters and fish habitats defy prediction

ICLARMs spectrum embraces artisanal fisheries, mariculture and fresh water aquaculture. To increase fish production research is dedicated to germplasm enhancement, breeding and genetic modifications, conservation of productive species, production technologies, nutrition, reproduction and pathology. ICLARM has designed computer models for stock assessment and several information and data bases for accession and dissemination to partners in its cooperative networks.

An issue critical to all CGIAR centres, not least to ICLARM is to decide what of all that is needed from basic and applied research is best conducted in-house and what can as efficiently, and at less cost, be carried out by other established scientific institutions.

International Crops Research for the Semi-Arid Tropics (ICRISAT)

Over 750 million people inhabit the semi-arid tropics, ecologies characterised by erratic and unreliable rain-fall, frequent drought and soils fragile in structure and low in essential nutrients.

In 1972 the CGIAR established a research centre, with a headquarters in Hyderabad, India and a modest affiliated programme in Africa. The African programme later expanded, now with two sub-regional research stations, one close to Niamey in Niger, a second in Zimbabwe near to Bulawayo. ICRISATs responsibilies embrace the improvement of several cereal and food legume crops, all of importance to the diets of many people resident in semi-arid tropical regions in Africa, Asia, the Near East, the Caribbean and Latin America. The crops in question include: (1) Sorghum, of which ICRISAT has collected and taxonomically identified over 30 thousand genotypes; (2) Millets, primarily Pennisetum (pearl, bullrush) millet; (3) Pigeon pea; (4) Chickpea, a responsibility it shares with ICARDA; (5) Groundnuts, a subsistence food and commercial oilseed.

In general, by various mechanisms, the ICRISAT crops display greater resistance to drought stress than cereals such as rice and wheat, and many other food legumes. ICRISAT improved types of Pennisetum millet are being grown in Pakistan, Sudan and Zambia. Sorghum hybrids developed for Sudan yield over 800 kg per hectare without irrigation compared with under 300 kg/ha from local varieties. Over 30 high-yielding types of sorghum from ICRISAT are grown in various countries of Africa, Asia and Latin America. Several African food processors and brewers have replaced imported barley by high-diastase sorghum to produce malt.

ICRISAT is a repository for native species and wild relatives of its mandated crops, cooperating with others such as the USAID INTSORMIL program, in expanding and diversifying its germ plasm collections.

ICRISAT leads the CGIAR Desert Margins Initiative (see pp. 130, 218).

Other CGIAR activities

The Group recently admitted the International Centre for Research in Agroforestry (ICRAF) an organisation created by IDRC with headquarters in Kenya. ICRAFs first purpose was to improve efficiency of land use and related resources, by integrating trees and other woody species into farming systems. ICRAF undertakes ecological diagnosis, improvement of multipurpose trees, and recovery of degraded land (Huxley 1983).

The International Service for National Agricultural Research (ISNAR) helps developing countries to strengthen their agricultural research, management, facilities and services.

The International Food Policy Research Institute (IFPRI) studies how food and agricultural policies influence food production and security. IFPRI has shown that high yielding crop varieties benefit large and small-scale farmers, and that producers of export crops enjoy a higher standard of living than those who raise only subsistence crops.

IFPRI has published extensively on policy related to agriculture, nutrition and food security; on poverty and disposable income; household surveys; on water rights related to irrigation; on income related to changes in eating habits; on livestock production and future demand for feeds and supplements; on the possible consequences of the Uruguay Round of GATT agreements for developing countries.

The CGIAR Secretariat at the World Bank in Washington and each of the CGIAR centres issue annual and interim reports of activities and results, reports readily available to interested scientists and serious students.

The CGIAR family of Centres and activities has made unique progress in agricultural research for the benefit of developing countries. Not least among the benefits delivered are the thousands of scientist who have been trained. Training courses include (1) post-graduate training for higher degrees, each of the Centres having established a formal relation with one or more universities; (2) practical grass-roots training in enhanced and advanced agricultural research techniques and technologies.

CGIAR Research Allocations and Priorities

In December 1994, the CGIAR published proposed research allocations upto and including Fiscal Year 1998 (TAC/CGIAR 1994a). The actual and projected financial requirements for the CGIAR system as a whole are tabulated below in current and 1992 U.S. millions of dollars.

	1993	1994	1995	1996	1997	1998
Current $M USD	230.6	247.5	265.3	284.4	305.1	328.6
1992 $M USD	221.8	228.9	236.0	243.8	252.0	260.1

In constant 1992 dollars, the CGIAR suggests an increase of only 17% over the 6-year period, a modest rise considering the expected demand for progressively higher production from an arable land base which, per capita, is falling on every developing continent. To raise output, while conserving the fertility and enhancing the productivity of the declining land base, requires significantly greater investments in research and related services than are proposed, both in the CGIAR centres and national agricultural systems.

In a companion publication (TAC/CGIAR 1994b) the proposed priorities and strategies for the CGIAR system and its family of centres are reviewed. To arrive at the priorities, the options and alternatives were analysed according to three "dimensions": (i) Activities, (ii) Spatial, (iii) Products.

The Activities dimension embodied:

- Conservation of biodiversity and other natural resources
- Breeding to improve germplasm
- Management of production systems
- Socio-economic and policy research
- Institution building including human resource development

The Spatial dimension covers:

(A) Four developing geographic regions:
 (i) Asia and the Pacific
 (ii) Sub-Saharan Africa
 (iii) Latin America and the Caribbean
 (iv) West Asia (Middle East) and North Africa.

(B) Nine agro-ecological zones, ranging from Warm-arid and semi-arid tropics, to Cool sub-tropics with winter rainfall.

The Products dimension includes crops, livestock, fisheries, forestry.

The Priorities publication describes how producing more food will increase pressures on natural resources, calling for more research on resource management and the conservation of genetic diversity. For the potential benefits of international research to be realised, national agricultural systems and human capabilities in Third World countries must be significantly strengthened. The CGIAR centres' record in the training of agricultural scientists and technologists

is impressive, more than 25,000 having been trained between 1985 and 1989, of which over 2000 were at post-graduate levels: Post-doc, Ph.D., M.Sc.
More specifically, the research priorities proposed include:

Natural resource and ecosystem conservation
- Ecosystems analysis; mapping of aquatic and terrestrial ecosystems
- Biology and ecology of useful organisms and pests; distribution and dynamics of food plants, animals and fish; of pests that affect them
- Land resource conservation; maintenance of productivity in relation to edaphic, climatic, hydrological and biological resources
- Aquatic resource conservation; monitoring of population dynamics

Germplasm enhancement
- Crop breeding to incorporate primitive and novel germplasm
- Breed improvement in livestock and aquatic animals
- Selection and breeding of multipurpose trees and shrubs
- Application of molecular biology to genetic mapping and genetic markers to assist crop and animal breeding programs

Production systems
- Socio-economic and technical constraints to sustainable production
- Design and assessment of novel and modified production systems
- Cropping systems, to include plant nutrition, plant protection, seed production and certification, post-harvest technologies
- Livestock systems - nutritional status related to feed and fodder; animal health and reproductive biology
- Arboreal systems - silviculture, plantation and natural forest management; nutrient uptake and cycling; economic control of diseases
- Aquatic systems - fish reproduction, pathology and nutrition

Priorities are proposed for Socioeconomic and Public Policy research including market and trade analysis, and cost-benefit studies related to adoption and consequences of technological research and development results; also for Institution Building to include training, counselling on organisation and management of research; information documentation and dissemination.

The above selected main headings of the research categories proposed in no way indicate the scope and diversity of the laboratory and field research implied; of the physical, human and financial resources needed; of the time required, under most favorable circumstances, to progress through various stages of research, demonstration, on-farm testing to eventual adoption and realisation of economic and social benefit. Agricultural research and development cannot be assessed in terms of sudden impact; the processes and products evolve slowly. In countries with advanced agricultural institutions, it takes at least ten years to bring a novel cultivar from early breeding to a stable registered variety.

FAO adopted as its motto "*Fiat panis*". The CGIAR and all whom its centres seek to serve might consider as their guiding motto: "*Festina lente*".

PESTS AND DISEASES

Pests that infest, infect, predate and are parasitic on growing crops include weeds, arthropods (insects, mites and spiders), bacteria, fungi and viruses. Pests inflict various insults and injuries: nematodes attack roots; insects and larvae damage vegetative and reproductive parts, some being carriers of pathogens; soil and airborne bacterial, fungal and viral infections cause debilitating diseases, some that are harmful to humans and animals; competitive and parasitic weeds deprive crops of nutrients; birds and rodents take toll of crops in the field and, together with insects and pathogens, after harvest. Plants and pests exist in dynamic relations one with the other. Pests mutate to evolve tolerance to pesticides to which they once succumbed. Flying insects and microbial spores are carried long distances by air currents. Exotic pests are introduced into foreign territories on imported commodities and on germ plasm that evades quarantine inspection.

Pests are controlled by various defensive measures: by plant breeding and genetic modification to enhance intrinsic resistance; by multiple cropping and other cultural practices; by biological control: release of organisms that are enemies of particular pests; by allelochemicals and biocides present in or extracted from other organisms; by chemical pesticides repellent or toxic to specific or a spectrum of pests.

Intrinsic Resistance

There is an extensive literature on genetically controlled plant resistances to insects and other pests (Philogene 1981). How a pest is attracted or repelled depends on the plant's physiology, its physical and chemical structure and composition. Volatile substances emitted by plants may attract (kairomones) or repel (allomones).

Pests can be repressed by genetically controlled resistance induced by conventional plant breeding: crossing a susceptible genotype with one that displays resistance to particular pests; or by techniques of transgenesis that transfer resistant traits from wild to cultivated species. Intrinsic defence systems can be stimulated chemically or by inoculation with a mild or avirulent strain of a particular pathogen.

To the lay mind, "biological" may appear less malign than "chemical". Biochemistry and structure determine if a plant is digestible and nutritious; easy or not to invade; suitable or not as a repository for a parasite's eggs. A plant's capacity to resist or repel a pest depends on its chemical composition: on the metabolites that accumulate in the plant's vegetative and reproductive organs. Resistance to insects may rely on physical characters such as tissue toughness and fibre content or on defensive chemicals, important among which are various phenols and alkaloids. Most simple phenols are extremely toxic; macromolecular polyphenols react with and render proteins insoluble, thereby reducing nutritional quality directly or indirectly by the immobilisation of digestive enzymes. Phenolic substances can deter and repel phytophagous insects that feed on plants by biting, boring or sucking on essential fluids. Some alkaloids interfere with insect development and survival. Neem (*Azaridachta indica*) synthesises a triterpenoid azaridachtin, a few parts per million of which can repel and restrict insect growth and be fatal to larvae (Prior 1989; IARI 1993; Philogene 1981, 1988).

Resistant traits include repellent characters that discourage insects from taking up residence or laying eggs; antibiotic characters that are toxic or noxious to particular pests; tolerant characters which enable plant growth to continue despite injury from or invasion by one or more pests.

Using modern transgenic tools plant breeders can transfer genes from unrelated organisms into new crop cultivars. The International Rice Research Institute is adapting molecular and cell biology techniques to achieve genetic improvement of cultivated rice species. Of more than 20 identified species of the *Oryza* genus, only two are cultivated as food crops. Some wild *Oryza* species carry resistances to pests and pathogens to which cultivated rice genotypes are susceptible. A prime objective of the IRRI program is to transfer genes for particular resistances from the wild to the cultivated species and thereby reduce dependence upon chemical pesticides.

Research on transgenic techniques to transfer resistances to pests from wild to cultivated species is a high priority in plant breeding institutes. Resistance by transgenesis is not, however a global benign panacea. Pests evolve means to overcome intrinsic chemical resistances by which they are for a time deterred. Instances of wide intergeneric crosses are recorded where the hybrid progenies, had they been released, could have caused insult to human health. In Britain and the USA wild Peruvian potato types were crossed with national standard varieties to improve pest resistance. The resistant traits transferred carried high levels of a neurotoxic glycoalkaloid, discovered by good fortune before the cultivars were released to farmers (H&W Canada 1989).

Historically, new cultivars were derived by breeding from cultivated species familiar as human food. As wild species, unknown in human diets, are more widely used as parent stock for hybrids, possible insults to health become less predictable. It would appear prudent that agronomic screening of novel wide-cross hybrids be subject to protocols that ensure wholesome nutritional quality and safety; that plant breeders and agronomists cooperate with nutritionists and toxicologists before they release novel wide-cross hybrids (H&W Canada 1989).

CULTURAL PRACTICES

Pest control by cultural practices works by disrupting a pest's preferred or established habitat and/or disturbing its life cycle. Most examples described are illustrative of particular conditions for many of which research is needed to clarify the underlying principles and to exercise systematic control.

Multiple Cropping Systems

Most of what is published on multiple cropping systems relates more to agronomy, to yields and productivity, than to pest control. "Multiple cropping" (MC) describes intensification by growing more than one crop on a given piece of land (e.g., a farmer's field) in the same year. It contrasts with monocropping in which a single crop is grown year after year. "Relay or Sequential cropping" describes growing two or more crops in sequence in the same year. In "Intercropping": two or more crops are grown simultaneously. Intercrops may be mixed without any orderly spatial arrangement; planted in evenly spaced rows or in strips wide enough to permit independent cultivation; in relays, the second crop is planted when the first has reached a particular stage of maturity. "Rotational cropping" is a repeated ordered sequence of successive crops, into which a period of fallow may be interspersed, a rotational sequence being spread over several seasons or years. In "Agro-forestry" woody species are grown in various patterns with agricultural and/or horticultural crops (ICRAF 1991; Huxley 1983).

Insect damage is reduced by species diversity of multiple cropping systems. Row intercropping imposes spatial and biological barriers that restrict movement of insects between plants. Rotational cropping restricts carry-over between seasons. In Amerindian agriculture maize was intercropped with beans and squash, the beans climbed around the maize stalks and the squash spread over the ground to shade out weeds.

Weeds

"Weed" derives from an Anglo-Saxon word "*weod/wiod*" interpreted to mean a herbaceous plant not considered valuable, useful or beautiful. Italian ("*malerba*") in common with other Latin languages defines weeds as "evil grasses". Neither is wholly satisfactory since one farmer's weed may be another's forage. In general, a weed is a self-propagating plant growing in competition with cultivated crops. In herbicide-free cultivations, weed control is among the most labour intensive on-farm activities. Competitive and parasitic weeds deprive cultivated crops of essential nutrients and reduce yields. Association of particular weeds with particular crops is not unusual. Crop rotations can disrupt weed reproductive cycles. MC systems can include species that compete effectively with or are allelopathic to particular weeds. Pest control by MC systems appears mostly empirical, research being needed better to understand and apply the biological principles involved.

The economics of weed and other pest control by cultural practices is difficult to determine particularly in multiple cropping systems. In principle there is cost benefit when the cost of control is less than the loss incurred by reduced crop yield from uncontrolled pests. The arithmetic is much easier with pesticide control of mono-crops than with cultural control in multiple cropping systems. Consequently, it is oftentimes difficult to demonstrate to farmers the advantages of changing from heavy dependence upon pesticides to cultural control.

Baker/CASAFA (1991) lists 48 broadleaf and 19 grassy weeds resistant to herbicides. Triazine is most frequently reported. Cross resistance to several chemically distinct herbicides was first reported in annual rye grass (*Lolium rigidum*). More instances of cross resistance are coming to light as weeds are attacked by a variety of herbicides.

BIOLOGICAL CONTROL

Biological control results from a pest's being attacked by a living natural enemy: a parasitoid, predator or pathogen. Though control of pathogens and nematodes is recorded, more effective are insect enemies employed against insect and weed pests, microbial pathogens being more difficult to disperse and maintain at epidemic levels (Greathead 1987, 1989).

"Classical biological control" introduces, propagates and releases exotic enemies. BIOCAT the data base of the International Institute of Biological Control records that during the last 100 years more than 4000 exotic enemies have been released to attack insect pests, over 500 to assault various weeds. At least 300 insect pests and 50 pernicious weeds are said to have succumbed. Notable results were in the USA (most in Hawaii), Canada, New Zealand, Bermuda, Australia, Fiji, Mauritius and India (Greathead and Waage 1983; Fewkes and Greathead 1978).

Effectiveness of biological control depends on crop value, the nature and number of troublesome pests, availability of critical resources, especially trained scientists and technicians equipped with facilities for diagnosis, operation and monitoring. Essential, but all too scarce, is taxonomic ability to identify correctly the pests and their specific enemies; to propagate and release the enemies in adequate numbers and patterns; to monitor progress and protect benign organisms. Because of the time taken for released enemies to reach adequate numbers, greater success is reported with tree crops, such as citrus, coconut, other perennials, and with vegetables in greenhouses, than with grains and annual field crops. Other patterns of biological control include conservation of natural enemies by creating environments conducive to their proliferation and activity. Cultivation of cover crops and heaping of farm residues can provide habitats for indigenous parasitoids and predators that are effective enemies of unwanted pests.

Augmentative control, which propagates and releases natural enemies, may be inoculative: seeding cultivated crops with natural enemies in greater numbers than occur normally; and/or inundative: the release of numbers sufficiently large to cause rapid reduction of the pest. Augmentative control can prove difficult in practice since release in sufficient numbers must be timed for when the crop pest is at its most vulnerable stage. Propagation and release in the numbers and at the critical time necessary, requires fecund organisms that can be produced economically on factitious hosts or artificial media. Augmentative control is generally best applied to crops planted successively over large areas. Showing promise experimentally is the propagation on artificial media of microbial pathogens prior to their being sprayed as quick acting biocides to control selected weeds, caterpillars and grasshoppers. Mycoherbicides, fungi pathogenic to particular weeds but non-toxic to benign organisms, have limited use since being specific they leave many weeds untouched. They also require long periods of contact to be effective (Greathead 1988).

Biological systems of control will expand and diversify using modern methods of genetic modification, though it will take considerable time and research for biological control to eliminate need for chemical pesticides. As with all pest control systems, the consequences of miscalculation are rarely predictable. High levels of professional skill particularly in advanced taxonomy will be needed reliably to devise, manage and monitor effective systems of biological control. Among many constraints, most evident is a serious shortage of suitably trained people, a deficiency in urgent need of correction in all regions of the world (Waage 1991; Waage and Greathead 1988).

PESTICIDES

Chemical Pesticides

Dekker (1991) cites estimates of the global expansion in pesticide sales. Between 1960 and 1989 (given in $M USD) fungicides increased from 340 to 4490, insecticides from 310 to 6200, herbicides 170 to 9500. Greatest use of herbicides is in the USA ($3.4B), of insecticides the Far East ($2.3B), of fungicides Western Europe ($2.1B).

The first selective organic herbicide 2.4-D was reported in 1941 and has been extensively used to control broad-leafed weeds in farmers' fields. A related phenoxyacetic acid herbicide 2.4.5-T was combined with 2.4-D in "Agent Orange" and used as a defoliant by the U.S. forces in Vietnam. Adverse effects among veterans of the Vietnam war, caused by an impurity dioxin, a persistent highly toxic substance, led to a ban on 2.4.5-T though the herbicide can be synthesised free from dioxin. Weed control with sulphonylurea herbicides seems now to be expanding. New advances in pesticides may include genetically modified biocides and viral agents that attack specific insects. Photodynamic herbicides, amino acids that disrupt a weed's biochemistry, are pest specific and rapidly degradable (Parmal and Devakumar 1993; Prior 1989).

In the USA, herbicides comprise 65% of all on-farm pesticides. The U.S. NRC reports herbicide treatment of 95% of all maize and soybeans; though variable among regions, farms and crops, pesticides make up 20% of all purchased input costs (NRC/NAS 1989).

With present knowledge, many plant pathogens are difficult to control without resort to fungicides. Fungal pathogens are ubiquitous, they are carried between continents by upper air currents, and they mutate to overcome intrinsic resistance in the crops they infect. Some, such as ergot and mycotoxins, impart acute and/or chronic toxicities. Biological controls are not notably effective against most fungi (Dekker 1991).

Gianessi (1993) discusses alternatives to chemical pesticides and points to the exceptional scientific and financial obstacles entailed in chemical-free farming. Non-chemical alternatives available are limited and, according to a Texas A and M University study, crop yields in the USA would drop markedly if such alternatives as exist were substituted for chemical pesticides. It is estimated that soybeans harvested would fall by 37%, wheat by 24%, cotton by 39%, rice by 57%, groundnuts 78%, maize by 32%.

There appears to be no clearly defined plan to make possible the abandonment of chemical pesticides. Nor has it been decided which of the many registered pesticides should be banned, and what alternatives are foreseen. Chemical pesticides have been cost effective and highly successful in controlling thousands of agricultural pests. To eliminate or substantially replace chemical pesticides in the USA, an adequately financed research program, with clear objectives, involving both public and private interests and competence must be set in motion.

At present, EPA assesses risks and benefits entailed in the use of particular pesticides, any of which may be banned or restricted if the risk entailed is unacceptable. During FY 1993 USDA spent about $134 M on pest management projects, most directed to reduction or replacement of pesticides in use. EPA estimates that with an absolute ban on ethylene bisdithiocarbamates (EBDC) fungicides the annual loss to vegetable and fruit growers would range between $270 M and $560 M.

Biological control methods are generally aimed at a single species. In contrast, many chemical pesticides cover a broad spectrum. In the USA as many as 100 crops must be protected against several thousand pests; over 200 active pesticide ingredients are used to attack them.

In general, Gianessi contends, chemical pesticides applied to fruits and vegetables pose more serious hazards than those sprayed over large areas of field crops. Many older pesticides were registered before the potential hazards were identified. Some time ago EPA set out to examine the potential risks of all active ingredients in pesticides registered before modern testing methods came into use. Over the past 5 years chemical companies have voluntarily dropped registration of close to 30 active ingredients and some 5000 pesticide products. Present cost of developing and registering a new chemical pesticide is of the order of $100 million. Consequently it is uneconomic to develop for other than a very large market. In the USA slightly more than 200,000 ha are devoted to apple orchards, 100,000 ha to lettuce, 20,000 ha to strawberries. These pale into insignificance against 30 million ha under wheat, 29 million under maize, 24 million under soybeans. Chemicals to control these large area crops are economically more attractive than for horticultural crops on which pesticides pose a more serious hazard to health (Gianessi 1993).

Pesticides with Biological Control

Biological control has many advantages but calls for diligent and competent monitoring to avoid unexpected misfortunes. Species released for biological control may have caused extinction of many other benign organisms. Thus, parasitoids and pathogens intended to control unwanted crop pests may themselves evolve as pests. There is little regulation of biological control and too little systematic analysis of overall consequences. Microbial pesticides: viruses, bacteria or fungi, are covered by registration requirements similar to those for synthetic chemicals. Prior to registration it must be demonstrated that the biocide poses no threat to human or animal health. Once a pathogenic virus is set loose it would be near impossible to recover or control it.

A total ban on chemical pesticides could cause serious impairment of crop yields particularly in tropical regions. Chemical control of pests is critical to economic production of high value horticultural crops intensively cultivated in hot humid climates (Philogene 1988).

Based on acute oral and dermal toxicities to rats, and on conditions of use, WHO has classed pesticides into five categories from "Extremely hazardous" to "Acute hazard unlikely". Most serious injuries to health result from misuse by farmers uninformed of hazards, or who receive instructions for use in languages foreign or unintelligible to them. Contamination of water sources and residues in food crops also raise health and environmental concerns. Difficulties in international control arise from inconsistencies in pesticide legislation among many jurisdictions aggravated by inadequate monitoring facilities particularly in many poorer countries (CASAFA 1991).

Most natural enemies are specific to a pest genus or species, rarely to a broad spectrum of genera. Consequently, cultivations that provide hosts and habitats for many pest species often require complementary integration of natural enemies with pesticides. For such conditions several pesticide options may be considered:

1) Broad spectrum pesticides
2) Specific selective pesticides
3) Introduction of natural enemies resistant to pesticides used.

Selective pesticides suppress specific pests leaving genetically unrelated natural enemies unharmed. Pesticides with growth regulators as their active principle may, for example, interfere with an insect's moulting behaviour without harming its predators or parasitic enemies. Other compounds inhibit egg-laying or cause infertility, preventing reproduction and propagation (Greathead 1988, 1989).

Selective microbial pesticides offer particular promise, of which strains of *B. thuringiensis* (Bt) provide a practical example. Transgenic techniques make possible the transfer and expression of the Bt toxin in tomatoes and other food crops. To counteract evolved resistance to Bt in particular pests, research is in progress to use concurrently several rDNA modified Bt strains, each effective against different susceptible pests. Given sufficient research support, other selective microbial pesticides could be identified and, if necessary, genetically modified to improve their effectiveness (Prior 1989; Waage 1991).

Insult and injury to natural enemies can be mitigated by methods of pesticide application. Systemic pesticides applied as seed dressings are unlikely to damage natural enemies that inhabit plant surfaces. Pesticides can be sprayed at times when natural enemies are least active. Pesticides put in traps primed with attractants will dispose of flying insects without interfering with pedestrian natural enemies.

Industrial and agricultural economics favour pesticides effective across a broad spectrum of crop predators and parasites, substances to be synthesised and distributed in large quantities. Pesticides that are highly selective and more safely integrated with natural enemies are likely to be demanded in quantities insufficient to be commercially profitable. To combine compatible chemical with biological control, requires substantial investment in systematic studies of the physical and biological relations between crops and their principal pests; between pests and their natural enemies; the ecologies in which they coexist; safe and ecologically benign biological, chemical and physical means by which pests' habitats and life cycles can be disrupted (Waage and Greathead 1988; Dekker 1991; CASAFA 1991).

Integrated Pest Management

IPM is a catholic descriptor of a broad highly desirable objective. The general purpose is not to destroy all pests but to determine and achieve "an economic threshold": a pest population and level of crop injury that is economically tolerable. One of the earliest publications advancing the principles of IPM came from scientists in California during the late 1950's. Proposed was pest control based upon economic and ecologically benign integration of effective cultural, biological and chemical management systems (NRC/NAS 1989).

IPM is better understood as a principle than a precise methodology. Agronomic and cultural practices; agro-environments; socio-economic conditions; genetic make-up of the crops cultivated; flora and fauna present, all influence the design of effective IPM systems. Economic thresholds will differ among regions, communities, farming systems and individual farmers. The cost of pesticides used is easier to calculate than cost and benefit of a complex IPM system. Estimates of IPM on-farm adoption and benefits vary among reports. One states that an investment of $10 million in a U.S. on-farm research program realised $500 million in benefit from reduced pesticide use without uneconomic drop in crop harvest. Others estimate less than 10% of U.S. cropland under IPM and suggest that efficient IPM may lead less to reduction in pesticide use than to more systematic sequencing of pesticide applications.

Safe and reliable IPM development requires dependable diagnosis of the relations between the crop(s) to be protected and the pest(s) to be controlled. Diagnosis should lead to design of control systems to be tested and evaluated first experimentally, later on-farms to determine dependable assessments of risk and expected benefit (NRC/NAS 1989).

Comment on Pests and Pathogens

Given their extraordinary facility to mutate and evolve to evade and elude farmers' defensive actions, crop pests and pathogens pose a most formidable constraint to raising production, and to food safety and security. Research on, and actions taken to improve pest and pathogen control demand and deserve far greater financial support and general public understanding than they now receive. Advocates who insist upon an absolute worldwide ban on all chemical crop pesticides, seem unaware of, or unconcerned about the excessive pre- and post-harvest losses, the consequent decrease in food crop availability, and adverse influence on food security that would inevitably result. When John Wesley wrote: "To be ignorant of many things and to be mistaken in others is a necessary condition of humanity", he could well have been referring to contemporary attitudes to the modern control of agricultural pests.

CHAPTER 6.
POST-PRODUCTION SYSTEMS

The Brundtland Commission Panel perceived Food Security as all folk having access to stocks and flows of food and cash adequate to satisfy basic needs: "(Food) security refers to secure ownership of, or access to, resources and income-earning activities, including reserves and assets to offset risk, to ease shocks and meet contingencies". In other words, everyone need not be a subsistence farmer, but all who are not must possess the means to acquire an adequate diet; to which end there must always be stable food stocks close at hand, sufficient in variety and quantity and at affordable prices (WCED 1987).

For most nations this is a rational interpretation, since among the wealthy that produce crops surplus to indigenous need, farm workers are a small proportion, many more being employed in food processing and distributing industries. Over 40% of developing country people live in cities and towns. Dense urban populations cannot be sustained by field-to-mouth subsistence agriculture. Famine and hunger resulting from tempests, drought and other disasters bear witness to the urgent need to carry over food stocks from regions and seasons of abundance to those of scarcity; to protect, preserve and distribute food from farms and fisheries in conditions that are safe, acceptable and affordable.

The literature on sustainable agriculture has much to say about production, but gives scant attention to post-production systems: the various activities between farms and consumers. An IUFOST working group of scientists convened at FAO estimated that the CGIAR international agricultural research centres devote less than 1% of their budgets to post-production systems (PPS); that donor support for PPS is uncoordinated and mainly directed to transfer of technologies from rich to poorer countries; development agencies seem unaware that food security depends on integration of production with post-production.

Post-Harvest Losses

Much is written about post-harvest losses: various forms of physical damage, biological and chemical deterioration that occur during storage, processing and distribution. The sources, causes and methods of determining post-harvest losses are discussed in many publications (Adams 1977; Bourne 1977; FAO 1977; GASGA 1978; Hall 1970; Harris and Lindblad 1978; Huysmanns 1981). Studies in Africa have shown how traditional farm storage systems, built of local materials, can be simply and systematically improved to be as efficient and more economic than donors' devices designed to be constructed with imported materials (IDRC 1981).

The West African workers applied the three principles essential in efficient grain storage: low moisture, low temperature, low oxygen. Sorghum and millet were air-sun dried; the threshed grain was fed into small silos before sunrise when diurnal temperature was lowest; the interstices among the grains were filled with wood-ash or cleaned sand (IDRC 1981, 1983).

Preservation by processing is achieved by control and modification of critical composition: of water activity by dehydration or increase in soluble solids; of ambient atmosphere by packaging under vacuum or inert gases; of pH by acetic or lactic fermentation; of temperature by chilling or freezing. Microorganisms or enzymes that cause spoilage or deterioration can be destroyed or inactivated by thermal processing or by exposing food to ionizing energy generated by high speed electrons from an electron accelerator or gamma rays emitted by the decay of a radioactive isotope. Preservation by irradiation is rare in not being of ancient origin. Though effective and safe within prescribed doses, it has suffered in the eyes of consumers probably from the stated initial purpose: "...to utilise waste products from the manufacture of atomic bombs" (Hayner and Proctor 1953).

Crop Composition and Utility

To utilise agricultural products efficiently, requires determination of biochemical composition and critical properties, and how these may be modified by processing. The major components: carbohydrates, proteins, lipids, and many thousands of minor substances in food raw materials vary extensively in chemical composition and in their response to processing technologies. Every organism synthesises chemical substances it needs for growth, health, reproduction and protection. Macromolecules determine functional properties, the utility of particular crop types for specific purposes: protein properties make some wheat types more suitable than others for breadmaking; starch composition, specifically the ratio of amylose to amylopectin influences the cooking quality of rice (Shewry et al. 1992; Juliano 1985).

Far less is known of the useful properties of the subsistence food crops of developing tropical countries than of food grains traded internationally. Plant breeders in developing regions seem devoted less to functional properties than with characters that influence yield. The range of diversity among crop genotypes held in international collections has been described. Diversity unused is diversity wasted. Opportunities for research on genetic diversity related to utility among Third World subsistence crops have been neither systematically explored nor exploited; properties critical to functional utility and how usefully to modify them genetically and by processing has received little investigation. Many thousand sorghum, millet and legume genotypes await the attention of food scientists and technologists.

Hulse (1994b) has reviewed research on legumes and the characters that influence functional utility of proteins and carbohydrates: active water properties; protein solubility and dispersibility; swelling and viscosity; gelatinisation; surface-activity in emulsions and foams; responses to electrolytes, pH and temperature; interactions with other macromolecules; technologies of fractionation and means to induce desirable biophysical modifications.

Future demand for meat and milk has been illustrated. Research on functional properties allied to the adaptation of known technologies of protein and carbohydrate modification could lead to substantial supplementation of meat and milk products by legume derivatives. Consumption of bread, other baked foods and pastas made from imported wheat is rising in Asia and Africa. By determining the properties of sorghum and millet genotypes compatible with those of imported wheat, African nations could develop composite flours and reduce their wheat imports. Discoveries of sorghum types high in diastatic activity has reduced dependence on imported barley for malting and brewing in Nigeria and Zimbabwe.

Food Quality

Food security requires that people have ready access to food that is adequate in quantity and quality. Literally, "quality" denotes unique properties that distinguish an object, material or product from all others. In the language of advertisers it implies a superior product. "Food quality" is not a precise designation, but is based on comparison of a food product against a defined set of standards or criteria, some quantitative, many subjective.

Governments' concepts of food quality relate to protecting their populations from insult or injury to health, from dishonesty or deceit. Food regulatory agencies lay down standards and methods of inspection to ensure that all foods sold are safe to eat and of the nature and substance expected. They prescribe levels of permitted ingredients, tolerances for unavoidable contaminants, and specify standards for weights, measures, packaging and labelling.

Food processors set their standards of quality, first to comply with government regulations, second to give their products recognisable and unique characters. Consumers' concepts of acceptable quality vary among persons and communities, and are influenced by ethnic, cultural, social and economic background and past experience. Consequently, there are no universal standards of food quality, though scientific criteria exist by which to assess some aspects of food safety (Curtis 1986).

Consumers' preferences and how they are influenced are of practical concern to food manufacturers and retailers, and of academic interest to social scientists

(Chisnall 1986; Mintz 1989; Scriven et al. 1988). Many studies indicate consumers' interest in food safety, nutritional quality and price. Concerns with environmental implications of raw material supply, with modern processing and packaging appear confined to a minority, modest in numbers but persistent in propagating their opinions.

Food Safety

Food security, food safety, human health and the environment are intricately interrelated. Composition of the air, water, machines, packages and organisms with which foods come into contact can affect their safety and stability. What may be considered safe by scientific methods of assay, may not coincide with what is perceived as safe by consumers. The precision and sensitivity of methods of analysis and assay are continually improving. Substances detectable several years ago in parts per thousand are now determined in parts per billion. Zero tolerance of a contaminant is, thus, not an absolute value but depends on analytical sensitivity (AFRC 1992; Garnick et al. 1988; Landegren 1988; Patel 1994; Hulse 1992).

It is generally easier to demonstrate what is acutely toxic or potentially harmful than what can be considered safe for all people at all times. Acutely toxic substances and potent pathogens are more readily recognised than what may pose hazards to health if consumed in modest amounts over long periods of time. It is biochemically true that one person's food may be another's poison. Mortality and morbidity from food and water-borne pathogens occur frequently where poor standards of hygiene and sanitation prevail. As might be expected, information and data on food infections and intoxications are more comprehensively available among the wealthier than the poorer nations, many of whom lack the facilities and resources needed to monitor diagnosed cases and to codify and enforce food quality and safety regulations. Records of food-borne infections and intoxications include only those diagnosed by professional services; sufferers who do not seek medical advice generally go unnoticed (Doyle 1994; Madden 1994; Food Tech. 1987).

Though statistics from any source are rarely complete or precise, the evidence is ample to illustrate that microbial pathogens are the most common cause of food-borne illnesses, greatly exceeding any caused by natural plant toxicants, environmental or other chemical contaminants. Educated extrapolations from confirmed cases estimate microbial food-borne sickness at several millions in Canada and the USA, the most frequent being salmonellosis, campylobacteriosis and staphylococcal intoxications. The economic cost, in terms of treatment, lost work time and other consequences is estimated in the billions of dollars. It is equally evident that incidences are on the increase and that many more originate from food service establishments than from food processed by well-established industries. Food service outlets, including street vendors that sell low-priced foods and employ poorly paid people of modest skills, are increasing in number in developed and developing countries. Consequently, food-borne microbial illness is more likely to rise than decline among urban communities (Doyle 1994).

Salmonellosis and campylobacteriosis, each causing close to 40% of food infections in North America, are most often associated with animal products. Others on the increase include listeriosis, the causative bacteria being often associated with rotting vegetation - one serious incident originated in salad vegetables fertilised with animal manure; and vibriosis (cholera), frequent sources being infected shellfish and molluscs in coastal waters close to dense populations. Vibriosis is more common in tropical than temperate regions, though infections occur from marine creatures caught along American and southern European coastlines during summer months (Madden 1984; Mossel 1989).

Nucleic acid probes, monoclonal antibodies and other modern methods of assay give hope of more rapid and reliable identification, diagnosis and control of virulent microbial infections. However, more research and investment are needed, particularly in Third World countries, if mortality, morbidity and economic loss caused by food-borne diseases are to be effectively controlled (Sharpe 1994; Stiles 1994; Mossel 1989).

Chemicals in Foods

All plant species synthesise substances and structures by which to discourage or destroy their predators. Various of these protectants are demonstrably toxic to humans and other animals. Toxicants produced by various microbes, plants and animal organisms are among the globe's most potent poisons. A purpose of plant breeding and crop domestication is to select genotypes low in noxious substances. Cultivated crops tend thus to be less resistant to pests, pathogens and predators than their wild relatives, entailing greater reliance on synthetic pesticides. Using modern transgenic techniques, plant breeders can transfer genes for specific pest resistance from wild into cultivated species. It may be questioned if adequate attention has been given to possible adverse effects on toxicological and nutritional properties of transferring biochemical resistances from wild plants, unknown as food crops, into cultivated species (H&W Canada 1989).

Natural and Chemical Pesticides

Though quantities are difficult to determine, it is highly probable that human diets contain more "natural pesticides" - substances that plants synthesise to protect themselves - than "chemical pesticides". U.S. observers estimate that Americans ingest at least 5000 different "natural pesticides" from their foods. Few of even the known "natural pesticides" have been scrutinised as thoroughly as have "chemical pesticides" (Ames 1983; Ames and Gold 1990).

Ames, Ames and Gold list toxins found naturally in foods, and others, such as mycotoxins, that result from microbial infections. They discuss the sparse attention given to potential carcinogenicity, mutagenicity (disruption of DNA), and teratogenicity (congenital malformations) of these many natural substances. To "natural pesticides" long present in diets, humans may have evolved biological detoxifying mechanisms. Though these mechanisms are normally broad rather than specific, they can be breached by exceptionally high or prolonged ingestion. Risks imposed by "natural pesticides", including any introduced by transgenesis, cannot be predicted without knowledge of chemical structures, concentrations, acute and chronic toxicities, and other related information demanded by government regulatory agencies before registration of new industrial "chemical pesticides" (Ames 1983; Ames and Gold 1990; IFBC 1990).

At least 600 distinct "chemical pesticides" are offered for sale around the world. In addition to national regulatory agencies, some 50 different international organisations are involved with pesticides. These include UN agencies, principally FAO, WHO, UNEP, ILO and UNIDO, who have recommended codes of conduct on sale, distribution and use. With so many organisations involved, diversity in attitudes and recommendations is inevitable and evident: what substances are permitted, which are restricted; residual limits prescribed for drinking water, harvested crops and processed foods. UN agencies have published a wealth of data on "chemical pesticides" but little on "natural pesticides". Several hundred substances are classified according to toxic hazard; maximum residue limits and acceptable daily intakes for close to 150 chemical compounds have been recommended. There is an obvious need for closer cooperation among international and national agencies that share interest in and responsibility for pesticide legislation and advice, taking account of both natural and synthetic substances (CASAFA 1991; H&W Canada 1989).

FOOD MARKETS AND MARKETING

The marketing of foods and other consumers goods is highly organised and vigorously competitive in the developed economies of Europe, North America and Oceania. Out of the neighbourhood grocery store, with its tradition of personal service to customers, have evolved the mammoth supermarkets and department stores where self-service is the rule, where shoppers make choices from thousands of brands and products. From being retailers of goods processed and packaged by others, grocery companies have aggressively moved into house brands and private labels in competition with their long-time suppliers. As the populations of Europe and North America increase more in average age than in numbers, competition among processors and supermarket retailers will doubtless intensify. Already, retailers' empires are transcending intercontinental boundaries and oceans as they seek self-protection and prosperity. Self-service denies to the grocer and processor direct contact and a lively awareness of customers' idiosyncracies and preferences, giving cause for immense investments in marketing research to try to predict what novelties will attract and satisfy thousands of remote potential customers (Chisnall 1986).

Harrison et al. (1987) review the markets and marketing services in countries of Latin America. They explain that development involves transformation from rural agricultural to urban industrial economies. As people migrate from rural areas to large towns and cities, and as urban centres grow in size, a rising proportion of the urban consumers' food expenditures are absorbed in marketing services, transportation, storage, product grading and processing. As incomes rise, so does demand for marketing services. As economies expand so do marketing services, in scope and importance as coordinators of food production, distribution and consumption systems.

Government and development agency investments have gone mainly to production resources; investments in marketing, transportation and essential infrastructures being at best a secondary consideration, if not overlooked entirely. Little investment has been devoted to credit, technical assistance or training to raise the standards of marketing sevices, whether controlled by private or public sectors. Development agencies appear to leave marketing services to evolve unhindered by any systematic government or private involvement. Various incentives encourage farmers to produce more extensively and efficiently; rarely are parastatal or private agencies given incentives to invest in and manage essential marketing and distribution systems. Yet food markets provide social and economic services, which need favorable government policies. Marketing services provide means of integrating production, distribution and consumption. Markets and marketing services are critical to national economies, to rural and urban food security.

Income and Food Expenditures

Estimates of mean or median incomes can be misleading in relation to demand for, and aggregate expenditures on, food. Income distribution among populations of Latin American cities tends to be highly skewed. In Recife, in north east Brazil, 14% of all families controlled half while the poorest 48% of families had only 14% of disposable income. Among all communities, equitable income distribution raises effective demand for foods in general, processed foods in particular, and for supporting marketing services. Among poor Latin American communities expenditures on food range from 50-70% of income. In Cali, a major city of Colombia, the lowest quartile devoted 82% of income to food. Within poor communities the income elasticity of demand for food is high. For every 10% increase in income, Latin American consumers spent 6% more for food and associated marketing services (Harrison et al. 1987).

Throughout history and from many recent studies it is evident that as poor peoples' incomes rise so do their expenditures to expand and diversify their families' diets. Among non-vegetarians, this inevitably results in rising demand for livestock products. In Latin American cities, where average incomes were below $500/year, between 30 and 50% of food expenditures was spent on meat, fish, poultry and dairy foods.

In the towns and cities of the developing world, food outlets fall into three broad categories: (1) public markets and street vendors; (2) neighbourhood stores, which can vary from a primitive shack to a permanent structure; (3) self-service supermarkets, often equipped with freezer and cold-store cabinets, which cater mainly to higher income groups. Neighbourhood stores and some street vendors, particularly those that dispense soft drinks and ice-cream, may have refrigerated cabinets.

There is a high degree of correlation between income level, the type of retail outlet patronised, and the quality of foods purchased. With the exception of soft drinks and beer, frozen and refrigerated foods are generally in greatest demand by higher income groups. In the Third World countries where general prosperity is rising, demand for beef, pork, poultry, fish and dairy products will certainly expand.

Throughout developing countries, public markets and street vendors are the main outlets for fresh fruits and vegetables, which, in tropical countries come in a wide assortment of varieties, qualities and prices. Climatic conditions combined with rough treatment during handling and transportation cause high levels of spoilage (Bourne 1977; NAS 1978).

Well-to-do North Americans may shop once a week at a supermarket, whereas poorer people in developing countries will buy daily what is immediately necessary from local markets. Among developing countries, supermarkets being few and far between, and public transport unreliable, poorer people favour local markets and neighbourhood stores.

Marked differences in attitudes and disposable incomes are evident among urban dwellers in the fast growing Third World cities. In cities such as Buenos Aires, Rio di Janeiro, Sao Paulo, Santiago, and a fast growing number in Asia and Africa, market outlets offering a wide range of frozen foods are by no means uncommon. Though among poorer people demand for refrigerated meat and fish is low, consumption of cold soft drinks, beer and, in many places, ice-cream is high and rising.

Fruit and Vegetable Marketing

Because of the wide differences in species and physiological state after harvest, marketing of fruits and vegetables is highly variable among and within tropical countries. Patterns and the means of reducing spoilage differ among different locations. For many in Asia, Africa and Latin America, bananas, plantains and such roots and tubers as cassava, sweet potatoes and white potatoes are dietary staples. Tomatoes and onions are also extensively popular.

Several Asian countries are home to large mechanised factories that extract cassava starch for export and convert the residue to animal feed. Cassava and bananas grown for export are cultivated from uniform varieties and are graded after harvest. For local consumption fruits and vegetables are grown by small-scale producers, with little grading before or after harvest. Consequently, responsibility for grading and quality control of local fruits and vegetables rests with retailers, who in many cities act both as wholesalers and retailers.

Poultry and Egg Marketing

Though many smallholders raise poultry, meat and eggs for their own consumption, in most Third World regions commercial poultry production is expanding. Broiler production is stimulated by the rapid growth of fast food outlets. Poultry meat, with a high income and price demand elasticity, is popular in most regions, being less constrained than pork and beef by religious interdictions

Dairy Products Marketing

Milk and its products are highly perishable and congenial media for transmission of many diseases. In more advanced nations milk is subject to public health regulations, for which to be effective milk must be collected, processed, transported and retailed through systematically controlled channels. Hygienic quality varies widely among and within countries, being generally satisfactory when collected from reliable producers, inspected and transported chilled to large city dairies. Milk consumed raw and unprocessed in villages can be a vector for all manner of chronic diseases. Even dry milk powder and infant foods, of which dry milk is an ingredient, can cause serious gastro-enteric disorders when reconstituted in unsterilised water from village wells.

In tropical countries milk production is often seasonal requiring that surpluses from seasons of excess are concentrated or dried, to be stored and later reconstituted during seasons of shortage. Dairies and collection centres often need heat exchangers and refrigerated storage smaller in capacity than is customary in large North American dairies. Nonetheless, several transnational corporations process and market dairy foods with modern facilities in Africa, Asia and Latin America. For the most part, they provide a useful service, their standards of quality being exemplary to smaller local enterprises. Several maintain research, product and process development laboratories in Third World regions, a primary purpose being to manufacture acceptable foods from local materials, thereby stimulating agricultural production, in some cases by contracts to local producers. Contract farming is gradually taking hold in developing regions, with food companies and breweries buying agricultural raw materials of specified quality in agreed quantities and guaranteed prices from local producers.

Red Meat Marketing

Except where proscribed by religious or ideological persuasion, meat from bovines and ovines; from hogs, avians and other non-ruminants is in rising demand across the world. In most major cities of the Third World cooked ground meat products: hamburgers, meat balls, kofta and sausages, are popular and major products of fast food outlets.

Meat is perishable and even in modern abattoirs it is difficult to prevent contamination of carcass meat by intestinal pathogens. Except in a few well controlled city abattoirs, hygienic standards are less than satisfactory, and wastage of potentially useful by-products is of egregious proportions in most Third World rural slaughter houses. Studies in Latin America show how animals are slaughtered during the night, carcasses are delivered unrefrigerated to retailers early in the morning, meat is sold to consumers later in the day (Harrison et al. 1987).

Higher operating standards and more effective use of by-products would reduce average prices, and increase gross availability of carcass and organ meats. As disposable incomes rise, demand for meat will also increase qualitatively and quantitatively, a trend further accelerated by the spread of fast-food outlets. Those owned and operated by large transnational corporations should, by example, serve to raise overall quality standards among local competitors. These patterns of growth in meat consumption should serve to raise hygienic and quality standards in abattoirs, transport vehicles and retail outlets.

Rural Marketing

A study in the Ashanti region of Ghana illustrates the conditions of food markets in rural areas of Africa (Hine et al. 1983) The Ashanti region covers 24,000 sq. km in which more than a million people live in 2500 small rural towns and villages. Most of the food sold in markets is produced in the region, the main crops being cassava, maize, cocoyam, plantain, various other fruits and vegetables. Ashanti markets also sell meat, fish, rice and tomatoes imported or brought in from other regions.

The Ghanaian food marketing system is largely under the control of women entrepreneurs who organise collection, transportation to markets, wholesale and retail sales. Produce may pass through as many as five intermediaries between the farmer and the consumer. Those who control the markets generally have an advantage over the farmer in fixing prices. Sales of each type of product are concentrated in an area of the market: all the yam sellers are clustered together in one area, all the plantains in another. Meat, fish and dry goods are each clustered together. Retailers of each commodity establish associations, each association functions as a price-fixing cartel. Each association appoints a leader, "the Queen". so that in each market there will be a Yam Queen, a Cassava Queen and a Plantain Queen.

In the region studied, the average distance between farm and village market is close to 4 km, most of the routes being footpaths. Access by truck is restricted to wider tracks and feeder roads, many impassable after heavy rains. Inutile transportation facilities between farms and markets account for excessive spoilage of meat, fish, perishable fruits and vegetables. In the study reported, least spoilage occurred where tracks and feeder roads were accessible to motor vehicles.

Though every marketing system of rural Africa, Asia and Latin America has its own peculiarities, the Ashanti conditions are indicative of the constraints imposed on food marketing by the inadequate attention and investment given to rural infrastructures and means of transportation.

Future Marketing Opportunities

To satisfy the expanding and diversifying demand for perishable food products in Third World countries the following are among many actions urgently needed:

- Construction, regular repair and maintenance of transport services, together with the roads, rails and waterways to carry them.
- Determination of electrical and other forms of energy needed for all urban purposes including those required by the food chain.
- Since public power grids are often overloaded, companies that process and distribute livestock, dairy and other perishable foods are wise to install auxiliary electrical generators. Small-scale operations may rely on ice and salt or, where available, liquid or solid CO_2.
- Investment credit for large installations must come from private and international development banks, many of whom seem unaware of the need for marketing services and the growing demands for perishable foods.

Marketing services in Third World countries are often inefficiently administered. Credit and other incentives are less readily available than for agricultural production. Contemporary wisdom that marketing should be left to free enterprise overlooks the fact that in most affluent nations government interventions, policies and regulations critically influence markets, and that subsidies to agriculture affect prices to consumers. Governments in most nations exercise control over highways, railroads, waterways, ports, power generation and distribution; taxation policies regulate fuel prices and transportation costs.

Agro-industries that process and market agricultural commodities as food and fibre, were the engines of economic development for every nation of Europe and North America. Integrated food marketing systems in Third World countries could provide employment for many rural and urban men and women. Effective marketing systems would increase food supply by reducing waste and spoilage. Efficient preservation systems would raise public health standards and reduce the immense economic and social costs caused by food spoilage and food borne diseases.

Free Markets

The 1993 UNDP Human Development Report proclaims the benefits of free markets which, it states, provide the most efficient mechanism for the exchange of goods and services by matching supply and demand, bringing together buyers and sellers, and so controlling prices that economies work efficiently. Free markets encourage human creativity and individual enterprise (UNDP 1993). The Report goes on to say that many markets are stifled by government regulations, particular interest groups which seek to favour those with greatest purchasing power or control of particular commodities and services. Markets that are not truly free but are prejudicially controlled can exacerbate poverty and unemployment. In the past, marketing of commodities has been often more exploitive than conservative, placing little value on the conservation of natural resources and the needs of future generations. Market prices by not reflecting a scarcity value may encourage depletion of natural resources.

Critical to food security is that all who need essential foods from markets possess the purchasing power to buy them. Serious constraints to the effective operation of many food markets exist in dilapidated or non-existent infrastructures, such as rural roads, transport vehicles and the means to protect perishable foods after harvest. Efficiently run markets are essential to food security. Markets, and the supporting services on which they depend, are critical to the safe delivery of foods to the fast expanding cities and towns of the Third World.

De Soto (1993) points to the higher prosperity of nations that have established free market economies and advances the hypothesis that markets function most effectively where properties rights are widespread and protected by law. In the USA legal property rights for land account for more than 40% of family assets; in Peru 90% of rural land property rights are informal and unprotected by legalised title. De Soto states: "Civilised living in market economies is not simply due to greater prosperity but to the order that formalised property rights bring."

The Universal Declaration on Human Rights states that equal access to property rights is a fundamental right of human kind. Among developing nations formalisation of property ownership and the legal right to buy sell and mortgage land is less advanced than in free market economies. De Soto contends that until the impoverished majority enjoy access to formal legalised property rights they cannot reap the benefits of a free modern market economy.

TRANSPORTATION

Energy in Transportation

Two basic physical principles favour railroads over road transports:

1. Rolling resistance between steel wheels and rails is less than between rubber tires and asphalt or dirt roads.
2. Aerodynamic resistance is less for railway vehicles since they move in long trains while road vehicles move singly. Consequently a railway vehicle requires less propulsive energy than a road vehicle of the same weight.

Efficiency of energy use depends on a vehicle carrying a full load. Railway wagons that travel full in one direction and empty on return are not efficient. Energy use efficiency increases with load, so that energy consumption to load size is a critical economic ratio. Railways are most efficient when carrying heavy loads. Railway energy efficiency measured per tonne per km is lower for light weight products such as cereals than for heavy commodities such as mined ores and fertilizers. Agricultural and food products, usually less dense and shipped in small consignments, are generally carried more economically by road.

In many countries the choice of options is severely restricted by existing infrastructures. Where dirt and gravel roads become impassable during rainy seasons, transport and distribution schedules cannot be wholly guided by conventional logistical and economic principles, or logistic-econometric models designed for North America or western Europe. In transporting refrigerated and other processed foods it may be necessary to design distribution chains that involve a hierarchy of storage points.

In several World Bank papers transport policy and planning, transport and economic performance, and road deterioration in developing regions are discussed (Bayliss 1992; Creightney 1993; World Bank 1988).

Transport and Economic Development

Only recently, it seems, have economic planners come to realise that transportation is not simply supportive of but critically catalytic to industrial and commercial development. Economic growth and distribution of goods and services can be seriously impeded in absence of efficient transportation within and between nations. Agricultural technologies and marketing services cannot expand beyond what are needed for basic subsistence without rural roads and other infrastructures that permit farmers to ship their produce to markets and to bring in the inputs their production systems need. Efficient transportation systems help bind together communities within a nation and nations within regions. In many developing regions transport has not been recognised for the critical role it must play in collective economic expansion, and the service it provides to all economic and social activities.

Creightney comments: "The World Bank has largely ignored the crucial role of transport in improving a country's economic performance and the well being of its population" Available evidence shows that many infrastructural interventions play a more important role than prices in achieving a balance between supply and demand. An efficient transport system stimulates production efficiency, helps to control consumer prices particularly as these are influenced by seasonal fluctuations. Publications on relations between transport and economic development in the Third World is sparse. There is little evidence from which to decide whether transport leads to growth or whether transport is developed in response to a demand from consumers and producers. Most often transportation facilities and passable roads are constructed where farmers and rural communities possess the power of political persuasion (Creightney 1993).

There is little persuasive evidence to indicate whether government infrastructural investment encourages or discourage private investment. Despite absence of systematic analysis, there seems little doubt that transport directly influences agricultural production and distribution systems. There exists an inevitable relation between transport and marketing services, and the costs of delivering goods between farms, processors and consumers.

Road Repair and Maintenance

In many Third World countries, the marketing of perishable foods is less constrained by the technologies of processing than by roads that are serviceable for transport vehicles. Throughout much of Africa and Asia rural and feeder roads that purport to connect villages and farming communities with urban markets centres are poorly maintained. Many, barely passable in dry seasons, become utterly impassable after heavy rains. Despite strong evidence of favorable economic returns to investments in rural roads and infrastructures, politicians appear less interested in spending on road maintenance than in construction of airports and conference centres in their capital cities. Perhaps it is easier to emblazon a Prime Minister's name over the portico of a new building than on truck loads of gravel or asphalt to fill potholes.

Poor road maintenance may raise the cost of transporting foods by 25 to 60%. It is therefore prudent in some instances to construct small or medium-size processing units, so dispersed that each is within year-round driving distance of producers and markets to be served, rather than to seek economies of large-scale in a single production centre. Studies in Asia and Latin America found most new industries located in areas close to major cities, thus depriving agricultural economies of the many advantages of preserving and processing foods close to places of harvest, and of opportunities for rural employment.

Bayliss (1992) describes how states of rural roads and related infrastructures exert a strong influence over where transnational agri-business companies decide to invest and establish processing and marketing operations. Another Bank study shows that in 85 developing countries, because of long neglect, over $40 billion is now needed for urgent road repair. The cost would be substantially less had the roads been repaired routinely. Deterioration of new paved roads begins slowly but accelerates rapidly after several years. Without early protective maintenance, road maintenance costs escalate to extremes. Even among affluent nations, governments are prone to neglect road and bridge repair until disintegration poses a danger to life and limb (World Bank 1988).

In the Third World road vehicles are being built faster than the roads needed to carry them. Consequently when contemplating investment in plant, equipment and/or technical services, it is wise to undertake a systematic study of the regions to be served: the infrastructures, road, transportation, energy and other essential resources available. Ideally, what are needed are mathematical simulation models that embody all relevant factors. Inter-related models are desirable for urban, rural, rural-urban, intra- and inter-regional transport, taking account of the products to be transported; traffic flows; conditions across zones and seasons of essential infrastructures; availability and reliability of vehicles and handling systems (Bayliss 1992). World Bank (1988) gives statistics of dimensions and states of repair of road networks in 85 countries of Africa, Asia, Middle East, Eastern Europe and Latin America.

It is estimated that 80-90% of all freight and passengers in Sub-Saharan Africa are carried by road. Sub-Saharan Africa's roads cover a total linear distance of 1.7 M km, roughly 1 km for every 4000 inhabitants, a ratio equal to India but much higher than in China. Because of neglect of repair and maintenance and abusive wear by heavy, overloaded vehicles, most African roads are in a sorry state of dilapidation. The World Bank estimates that $1.5B needs to be spent annually over the next decade simply to repair present damage and restore roads to a fit state.

Some improvement is promised by an African Road Maintenance Initative to be supported by a consortium of the World Bank, the European Union and nine European governments. In terms of reduced transport costs, more expeditious movement of goods, particularly perishable commodities, impressive economic gains may be expected from these long-overdue investments in the maintenance of Africa's roads (World Bank 1994; Economist 1995a).

CHAPTER 7. FOOD SCIENCE RESEARCH

TECHNOLOGICALLY ADVANCED NATIONS

The United States

A Research Committee of the (U.S.) Institute of Food Technologists has defined and described the priorities for food science research in the United States (Food Tech. 1993). The committee's report states: "Research is needed ...to optimize safety, quality and value" and addresses needs relevant to the 21st Century: diet, nutrition and health; biotechnology; environmental issues; food engineering and packaging; molecular basis for food functionality. Defence demands absorb 59% of total U.S. research expenditure; since 1970 defence research has increased while civilian R&D has declined. Of U.S. government investment in agriculture less than 15% was for research. "While the overall need of R&D investment has been...recognized recently, the importance of R&D for food science and food technology has not." (Food Tech. 1993).

Relevant to Food Safety the report calls for research to provide: Better methods to detect the nature and origins of food pathogens and toxins, and of reporting food borne diseases; improved toxicology tests; means to decontaminate infected raw materials; assessment of hazards and safety; costs imposed by foodborne diseases (*see* Roberts 1989).

The IFT recommendations for Diet, Nutrition and Health are consistent with those of the U.S. Surgeon-General and U.S. NRC Committee.

Biotechnology is expected to bring forth novel materials by genetic modification of plants and animals; identification of mechanisms that give rise to deterioration; new enzymes and biopolymers to enhance and expand ingredient inventories; immunoassays and DNA probes to serve as diagnostic tools for pathogens; better understanding of functional properties and modes of action of enzymes and macromolecules natural to and induced through genetic modifications (Food Tech. 1993).

In relation to environmental concerns the report lays emphasis on controlling waste and efficient use of raw materials. It proposes:

Better understanding of the properties of solid and liquid waste materials to permit their conversion to useful by-products; economic use and recycling of water by physical and biological decontamination; alternatives, such as supercritical fluids, to industrial solvents; development of refrigerants to replace ammonia and CFCs; adaptation of by-products to energy generation; reduced use of packaging material which, by extrapolation of present practices, will amount to 32 M tons per year by the turn of the century.

Research for Engineering, Processing and Packaging reiterates some of the foregoing and adds: Quantification and modelling of properties and processes; development and monitoring of on-line sensors to ensure correlation with desirable properties; heat and mass transfer studies across the spectrum of food processing technologies; development of processing systems that minimise modification of essential properties.

The Committee estimates results of research proposed could realise annual savings to the U.S. economy of *ca.* \$15 B, and recommends an added annual investment of \$150 M in food science research (Food Tech. 1993).

The food science research priorities recommended for the USA were formulated by a committee of scientists drawn from industries, academic and other scientific institutions. They are broad based, taking account of perceived developments and opportunities as they bear upon American food industries and the general public. Since the knowledge derived is unlikely to offer immediate competitive advantage to any particular companies, the research required could be carried out cooperatively by universities and industries, the results being published openly.

As is suggested later, a compelling case can be argued for the research needed to be carried out through international cooperation.

Another study by a U.S. scientific committee, though not specifically devoted to research priorities, bears important scientific implications for food industries and consumers. The International Food Biotechnology Council (IFBC) was established to identify issues and assemble a set of scientific criteria to evaluate and ensure the safety of food and food ingredients derived from plants and microorganisms by genetic modifications (IFBC 1990). The report cites variances in composition of traditional foods in terms of nutrients, naturally occurring toxicants and microorganisms. It describes techniques by which genetic changes can be brought about; methods by which to assess safety in modified food crops, in whole foods and complex mixtures, in foods derived from microorganisms. It discusses legal and regulatory issues and the U.S. Food and Drug Administration's disposition towards such issues as they relate to agricultural commodities, biological ingredients, chemical additives and food products derived from biotechnological processes.

Research by Food Processing Industries

Investments in research by food processors, even liberally defined, are small compared to those of pharmaceutical and chemical industries. Unpublished data assembled by the former Science Council of Canada indicates that of more than

3000 food processors registered in Canada, fewer than 80 were "research-active" in 1990, and that investment in research, on average, was less than 0.2% of total sales value. Among West European nations, the Science Council estimated average food industry research investment between 0.2 and 0.3% of total sales.

Though product innovation is sometimes proclaimed as the industries' vital driving force, it is probable, among even the biggest companies, that more is consistently gained by reducing operating costs through modifications and adjustments to established processing systems and technologies, than from launching new food products.

Costs and risks of bringing new products to profitable markets are largely unpredictable and failures are reportedly high, a fact of which food scientists in Third World countries should be fully conscious. Hollingsworth (1994) reckons the failure rate for new food products in the USA exceeded 80% during the decade 1984-1994, "hitting a peak of 91% in 1989". The longer the distance between producer and customer, the greater the difficulty and risk of predicting the acceptance of a new product. A local butcher or baker who sells directly to customers can maintain a more consistent and timely awareness of their preferences and predilections than a transnational corporation with centralised processing and products distributed through extensive networks to distant retailers.

The practical principles of most technologies of food preservation and transformation have been known for hundreds of years. Hermetic storage, milling and fractionation of cereal grains; expression of oil from oilseeds; panary, ethanolic and lactic fermentations were familiar to ancient Egyptians. Preservation by refrigeration, "boil-in-the-bag" (with seasoned ground meat sealed in the body cavity of a squid or a cleaned sow's womb), sausage-making, thickening of sauces by starch washed-out of cereal flour, were among technologies widely practised by Republican Romans.Unlike electronics, food technologies were discovered and developed empirically;only during the past century have scientists systematically studied the biochemical and biophysical principles upon which nutritional and functional properties depend.

Past progress was most evident in the replacement of human labour by machines driven by steam and internal combustion engines, or electric motors.Though first used to pump water out of deep coal mines, one of the earliest industrial applications of Watt's steam engine was to drive machines in a London flour mill.

As processes and inventories become more controlled by computers, so are laboratory technicians being replaced by automation of analytical methods; by diverse on-line sensors, instrumental monitors, computer image analysers that employ high resolution micro-photographics.

Scanning Tunneling Microscopy and Atomic Force Microscopy facilitate elucidation of structures and sites of reactivity of carbohydrates and other macromolecules. Scanning probe microscopies offer opportunities to probe complex macromolecular structures;to discover how biochemical interactions influence functional behaviour;the way specific enzymes interact with carbohydrates and proteins; means by which biochemical and biophysical properties of food components may be modified and tailored to serve particular novel purposes. Food science is gaining access to an expanding spectrum of techniques whereby to diversify and enhance industrial functional utility and nutritional qualities of a vast assortment of natural and synthetic biological materials (Hulse 1992).

To realise many opportunities critical to food security, cooperation among companies, universities and government agencies is essential. Hollingsworth (1995), citing no statistical data, states that following a decade of "downsizing and mergers", U.S. food industries now invest far less in research than was the case during the 1970's and early 80's. Since the market place for processed foods, and the opportunities for socially and economically beneficial application of food science are fast transcending national boundaries, there is a compelling case for cooperation on an international scale, research in which industries and food research institutions in both technically advanced and developing nations would collaborate. Among others, the British industrial research associations demonstrate how pre-competitive research can benefit a whole industry.International agricultural research results indicate the benefits that could be realised from global cooperation in food science.

The International Council of Scientific Unions (ICSU), together with several of its members and other international agencies, has inspired extensive cooperative research on global climate change and related environmental issues.As this is written, no comparable international cooperative program is dedicated to food security, with a program that comprehends both production and post-production systems; that includes food and nutrition research, agricultural and aquatic sciences, together with diagnostic and design studies on systems of safe storage and distribution.

If they were so willing, FAO in cooperation with the World Bank, with major food processors, their research and professional associations, could bring such an essential international programme into active existence.

Food Research in Developing Countries

In principle, as synoptically stated, several of the recommendations from the U.S. committees are of relevance to agro-industries in Third World countries. Food safety, as it is endangered by food pathogens, is critical in regions where water quality is suspect; where distribution systems lack adequate refrigeration or alternative reliable means of protection and preservation; where food-chain distances, spatially and temporally, are progressively extended as urban communities burgeon in size and complexity; where "fast-foods", made up of mixtures of ground meat or fish with other perishables, are dispensed by street vendors.

Nutritional qualities, wholesomeness and functional utility of novel and unfamiliar foods need to be systematically monitored whether their unusual compositions and structures arise from genetic modification by crop or livestock breeders, or from processing technologies. Capacities and resources of food regulatory agencies in the most affluent nations are often overstretched by the demand to legislate for and monitor the mass of innovations presented for approval by processors and suppliers of ingredients. In many developing countries, food legislative agencies, if they exist, are inadequately equipped and staffed. Infections carried in food, common in developed countries, could reach epidemic proportions among ill-protected urban communities in the Third World (WHO 1984).

Humanistic and ethical considerations put the onus on international agencies and affluent nations to ensure that Third World communities are not first to discover that certain novel foods may be detrimental to human health. Facilities, resources and procedures needed to evaluate genetically modified materials, as described by the U.S. committee (IFBC 1990), are not available in most developing countries.It is therefore for the affluent to demonstrate that novelty and safety always go together.

Research priorities applicable to all developing economies cannot be prescribed. Nonetheless certain common constraints are evident. Foremost is the need to preserve and process food materials more efficiently; to reduce waste and spoilage between places of harvest or slaughter and the consumer. Though donor and development agencies collaborate in support of research and training in production technologies, support for post-production research and development has been uncoordinated and in some instances conflicting. Some years ago, six agencies offered six different systems of grain storage to West African farmers, all reliant on prefabricated, imported materials. A West African research group demonstrated that on-farm grain stores, constructed entirely of local materials, when their design and methods of use were based on known scientific principles, were much cheaper and as effective as any the donor agencies proposed to import.

Rarely has the CGIAR discussed research that relates to post-production systems. Effective and economic utilisation of the extensive genetic resources available in international collections is yet to be realised (TAC/CGIAR 1994c). Research to explore how diversity in tropical crop germ plasm collections relates to functional utility is urgently needed; genetic diversity unused is diversity wasted. The international agricultural research centres need to cooperate with universities and food industries in order to relate genetic diversity to utility.

An international cooperative program, financed by a consortium of donors and industries should explore how the vast genetic diversity that exists could be utilised by industries in Third World countries to reduce dependence on imported materials and expand employment in rural communities. It could function by cooperation among existing research institutions, with an international advisory body to determine priorities, where and how they could best be pursued (Hulse 1992).

Mechanisation of artisanal food processes has long been a dominant feature of food technology development among affluent nations.It is therefore not surprising that machines are gradually replacing hand labour among rural African communities, women being relieved of the drudgery of hand-pounding grains by small self-contained mills.To advance cereal milling, research is needed to identify grain characters that are conducive to mechanical decortication, particle-size reduction and fractionation, and for plant geneticists to incorporate desirable functional properties into their breeding and selection programs. There would seem to be little advantage in achieving higher harvest yields of grains that are morphologically and structurally unsuitable for subsequent processing, or are considered unacceptable in quality by consumers.Breeding and selection of crops for properties demanded by processors, a *sine qua non* among prosperous nations, are objectives comparatively neglected in Third World nations except for crops to be exported. As food industries in Asia, Africa and Latin America expand and diversify to satisfy needs and demands of urban populations, so must plant breeders seek to balance such desirable agronomic traits as high and sustainable harvest yields and inheritable pest resistance, with properties that control commercial utility, nutritional qualities and consumer acceptance.

It is encouraging that regional food science associations are now established, for example, in Eastern, Central and Southern Africa, and among ASEAN nations. These, and their member institutions deserve and need support comparable to that given to international agricultural research. When FAO and the Government of India created an International Food Technology Training Centre at Mysore, facilities for food science research and technology development barely existed in Asia. Since 1964, the Centre at Mysore has trained over 4000 men and women from Asia and Africa. Now, university departments of food science and technology and national food research centres exist in many Third World countries. Unfortunately, most are inadequately financed and supported, with too little opportunity for productive activities. Some have been diverted by donors whose concepts of priorities bore little relevance to local immediate needs and opportunities. Others, happily have established working relations with local food industries to their mutual benefit.

The Director-General of FAO has declared food security as a major priority. As no other international agency is endowed with a comparable mandate, or scope of responsibility, it is hoped that FAO will devote the attention to food science and technology, and to agro-industrial development as are critically essential to food security in the Third World. Several divisions of FAO work in agro-industrial development. In some, industrial utilisation is integrated with production, marketing and distribution, in others such essential integration is less evident. In cooperation with the World Bank and regional development banks, with agricultural research institutions and agro-industrial organisations, FAO could provide the inspiration and leadership for a truly integrated food security program, one designed to embody all components essential to food security, a program designed to stimulate international cooperation in research and development along the lines suggested above.

CHAPTER 8. HUMAN RESOURCES: FUTURE REQUIREMENTS

The literature about social, economic and technical development is replete with such adjectives as "interdisciplinary", "integrated" and "holistic" indicating that "development", however defined, is a complex mix of many interdependent variables. Historically, universities have been organised into specialised faculties and departments, each devoted to graduating scientists and other scholars in discrete disciplines. Undoubtedly, future progress in agriculture, fisheries and their related processing industries will arise out of research by scientists of many disciplines and specialisations.

As essential but less widely evident, are systems scientists: men and women able to conceive how various interdependent components must be systematically unified. Scientists in agricultural research centres have dilated beyond their disciplines into farming systems research. In principle it bears similarity to market research in that it starts by studying and assessing the ecological, social, economic and technical environments in which a farming system exists; the resources available and the resources needed to adapt and adopt desirable new practices; the opportunities for and constraints to technological change and the risks likely to be incurred in consequence.

The International Council for Research in Agroforestry has devised a research process of diagnosis and design relevant to land-use systems in which various crops, livestock and arboreal species are combined to be economically productive while protective of natural resources. Agroforestry is basically a process of systems analysis.It illustrates the systematic evaluation and planning necessary to every nation's food security (Huxley 1983; ICRAF 1991; Swinkels and Scherr 1991). The systems methodologies of ICRAF, IRRI and other international centres deserve study by all who seek to formulate food security policies.

Systematic diagnosis and design processes, conceived and controlled by men and women able to comprehend complex production, preservation and distribution systems are essential to the realisation of every community's food security.There is little evidence that governments or development agencies are encouraging the formation of such people. Crop and livestock production specialists cannot alone provide for food security in any nation.

A ditty of yesteryear asked: "How're we going to keep them down on the farm after they've seen Paree?" The magnetism that draws young people from rural to urban habitats imposes severe constraints on food security. Fewer primary producers must provide for more non-producers; farm production must increase per unit of land and of labour; farmers must adopt more productive, unfamiliar technologies, to which end they need access to research and technical advisory services. Extension services, in most countries, are understaffed and inadequately financed. As youths migrate to cities, technicians available to rural extension services become evermore scarce. While services to farmers are far from adequate, technical and management advice to rural agro-industries in most nations is non-existant, food industrial research and development centres, where they exist, being most often found in large cities.

Societies in which women demonstrate a superior ability to manage money and markets are not unusual. Yet women have often inferior access to sources of credit, to formal education and training. Donor, development and government agencies could wisely invest more in training of women particularly for the management of agri-business in rural communities.

Food security, however defined, is for people and must be planned and brought to realisation by people of many skills and specialisations; men and women of unusual vision, imagination and training; people able accurately to assess present needs, to predict future needs, to conceive of all that must harmoniously brought together to satisfy these needs.

It is encouraging as this text goes to press to observe that the World Bank has devoted its 1995 World Development Report to "Workers in an Integrating World". The Report addresses development in relation to employment; the possible influences of international integration and expanding global markets on employment opportunities; the important role and responsibilities of governments as public policy makers and as employers; the prospects for workers of diverse skills in different regions of the world during the 21st Century (World Bank 1995).

It has been stated before, but deserves repetition, that development: technical, social and economic, however conceived or appraised, is for the benefit of people, and is brought to realisation by people of many attributes and skills. It is therefore upon people, not primarily upon science and technology, that development policies must concentrate. It is upon the needs and resources of people, especially the poorest in society that the vision of those who devise policies must be focussed.

EPILOGUE

It is accepted, as a fundamental precept, that all men, women and children are entitled to unconstrained access to food sufficient to sustain their bodily functions and activities. Opinions differ on what constitute adequate diets for people of different ages and conditions, and new prescriptions will appear and be modified as research throws fresh light on diet related to health.

Food security depends on diverse biological, physical and logistical processes, each conditioned by prevailing social, economic and political environments. Where humans do not intrude in the ecology, various other organisms will compete for the resources available, the mix and balance shifting dynamically with changing climate and as different organisms gain dominance. If by some mighty global cataclysm the human race were wiped out, its absence might be mourned by only a few body lice and intestinal parasites. But so long as men, women and children inhabit the planet, they have a basic right to perpetual food security.

In light of present knowledge and past experience, what hope is there that all who have need will enjoy future access to food security?

That many, particularly among poor communities, will suffer chronic or periodic food insecurity and malnutrition seems inevitable considering present political priorities with crumbling financial support for food and agricultural research and poverty alleviation. FAO has declared the pursuit of food security as a high priority. One hopes the Organization can assist governments to enact policies that conduce to sustainable systems of food security, specific to the needs and resources of their citizens, and that member governments will give to FAO the requisite political and financial support. To ensure future food security for all people will demand a more imaginative and conservative utilisation of essential available resources than at any time in human history.

There is inevitable discord between people dedicated to protection of endangered species and those whose livelihood depends on activities and industries that may violate ecological and species integrity. Acquisition of food and fuel by poor people in climatically stressed environments will conflict with ecological and resource conservation. Soil degradation is extensive in the drought-prone regions of Africa, the degradation being exacerbated by over-cultivation and over-grazing - raising crops and livestock without restoring soil fertility; by depleting arboreal and other vegetative cover to expose fragile soils to erosion. Such depredation, often neither wilful nor mischievous, will persist and spread if governments and development agencies do not act to eradicate the underlying causes. The Desert Margins Initiative, mentioned earlier, will pursue these kinds of objectives (ICRISAT 1995).

Predicted food demand rests on population growth, income increase and elasticity. Globally, rates of population growth are forecast to fall, with greatest numbers in Asia. While global food production has risen faster than population, food production per capita in sub-Saharan Africa has fallen continually over the past 20 years. Sub-Saharan populations are expected to rise threefold by 2025, by which time Asians will have raised their numbers from 2.6 billion to 4.4 billion (UN 1988). Arable land per capita will fall from 0.32 to 0.18 ha in sub-Saharan Africa; 0.17 to 0.12 in Asia; 0.36 to 0.26ha in Latin America (TAC/CGIAR 1994b).

Over many centuries, agricultural output rose by extending farm land. Since 1960, most increase came by raising crop yields (World Bank 1992). Yield increases resulted from: genotypes and phenotypes with enhanced yield potential - cereal plants with higher grain-to-straw ratios than their antecedents; extended use of fertilizers and irrigation; and effective eradication of crop pathogens and parasites. For how long food crop yields can continue to rise cannot be prophesied.

Hopes that induced genetic modifications will raise yield potential, improve crop resistance to drought and other environmental stress, and reduce dependence on chemical fertilizers, are yet to be widely realised. Induced pest resistance by transferring genes from wild species with long-evolved resistance into cultivated species, appears well advanced. Since arthropods and microorganisms that infest and infect crops can mutate to elude their host's biological defence mechanisms, resistance induced by transgenesis will require persistent monitoring and periodic genetic remodification. Valuable progress is reported in the genetic mapping of food crops using molecular markers to identify chromosome DNA sequences that control particular agronomic traits (CIMMYT 1995). Genome mapping is revealing how biochemical properties that influence nutritional quality can be genetically controlled (Butler 1994).

Research to increase some crop yields still progresses. The initial prototype of a new rice plant, with a 25% higher yield potential than established varieties, is reported from the International Rice Research Institute. Existing high yielding varieties push out 20 to 25 tillers (stem shoots) of which no more than 15 bear grains. The new phenotypes bear fewer than 10 tillers each with a panicle of 200 to 250 grains. Present varieties produce, per hectare, *ca.* 20 tonnes of biomass, equally partitioned between grain and straw, whereas the new types generate close to 12.5 tonnes of grain with 8.5 tonnes of straw (IRRI 1994).

By market value, close to 30% of the world's crops are harvested from irrigated land. Roughly 29% of Asian rice land, yet barely 3% of crop land in sub-Saharan Africa is irrigated. Sustainable productivity on African drylands would rise substantially if farmers had access to efficient irrigation, such as was successfully employed in desert reclamation in Egypt (El Lakany 1991). Of an estimated 4 trillion cubic metres that flow from African rivers, little is used for crop irrigation (Hulse 1993).

Yudelman (1994) states that between 1966 and 1988, capital costs for new irrigation systems in Asia rose by upto 160%/ha. African smallholders in drought-stressed areas are normally too poor to afford irrigation, fertilizers and other inputs to raise their agricultural productivity.

Adequate water and sustained soil fertility are essential to feed the people already alive let alone the 80 million new mouths expected in the Third World every year well into the 21st century. Two-thirds of these will be born in Asia where, since the early 1960's, farmers have consistently raised crop production, with 90% of the increase attributable to higher yields. Virtually all arable land in Asia is now under cultivation. Rising, shifting demands by Asian urban populations will call for higher local agricultural output and/or the financial means to import. Brown and Goldin (1992) forecast that cereal grain imports into Asia and other developing regions could rise from the present 65 million tonnes to over 100 Mt by the year 2000 and continue upwards well into the next century.

Rising demands on insufficient productive land, persuade Carruthers and Morrison (1994) that early in the 21st century developing nations in tropical and sub-tropical regions will find it economic to produce and export many of the world's manufactured goods, while the northern temperate countries will grow and export most of the planet's food.

Reform of the European Union's Common Agricultural Policy (CAP), and implementation of the GATT agricultural agreements, are destined to reduce direct subsidies, internal market prices, bound tariffs, exports of basic and processed foods (Mehta 1995). How changes to GATT and CAP may influence north-south agricultural trade, whether food commodity prices will become affordable to the poorer developing nations, and in what degree Carruthers' and Morrison's predictions will come about are neither certain nor predictable.

Overall, the world could produce food sufficient to satisfy the basic nutritional needs of more people than now exist. Chronic malnutrition and hunger could be redressed by equitable distribution of essential resources. But present political dispositions and parochial egotisms offer scant hope that the earth's resources will ever be equitable shared, so that little useful purpose is served by prophesying global carrying capacity: how many healthy people now, and 100 years from now, the planet can sustain. Practically, food security must be assessed for specific communities, taking account of their collective needs and resources.

How to provide and assure sustainable food security for millions of poor rural people will persist as an awesome and obstinate enigma. Of fast growing dimensions and bewildering complexities are the food and nutritional needs of expanding urban populations. From 1960 until the year 2000, the planet's city dwellers will have grown by 1.5 billion. By the century's end, 17 of the world's 21 cities that are home to over 10 million men, women and children will exist in Third World countries. While poor rural and urban persons share similar afflictions, constraints and disadvantages, the technological and logistical difficulties to be resolved to provide food and other basic needs for rapidly congesting and dilating urban conglomerates are far from being diagnosed, let alone resolved for remedy.

In most nations, the prosperous and powerful are rarely afflicted by chronic hunger. Malnutrition among opulent people more often results from excess than deficiency.

While millions in Africa, Asia and Latin America struggle for access to barely enough food to survive, corpulent and over-indulgent people in Europe and North America pay handsomely for products that fill but don't fatten. Fat replacements, substances with little or no caloric content but which imitate the physical and sensory properties of animal and vegetable lipids, are among the fastest growing of food ingredients. U.S. retail sales of "low-fat foods", in which natural lipids are replaced by modified proteins, celluloses, polysaccharides, gums and various other chemical synthetics, reached *ca.* $32 billion in 1991 and are predicted to rise to *ca.* $55 billion in 1996 (Jones 1995). Since, by definition, "Food is what is taken into the system to maintain life and growth", it is questionable if non-caloric, synthetic lipid imitations can rationally be regarded as foods.

Chronic deficiencies among poor people could be alleviated by clearer sighted government policies and more sensitive compassion by affluent nations. Both must recognise that sustainable food security demands integration of the means of production with the means of adequate storage, safe distribution and delivery.

Governments seem reluctant to formulate comprehensive food security policies, policies that take heed of changing demographic conditions, rising purchasing power and diversifying demand. They could be assisted to this end, if FAO in cooperation with IFPRI were to provide a food security policy diagnosis and design service, a service which enabled governments to predict future demand and to determine the resources and services needed to ensure adequate production integrated with safe and reliable systems of preservation, transformation, marketing and distribution.

Science can provide technological means; but it cannot promise global panaceas. Those with political power and financial resources might best be guided by the essential principle of food security as it is succinctly captured in the Book of Proverbs:

"Give me neither poverty nor riches; but feed me with food convenient to me".
(Proverbs 30, viii)

REFERENCES

ACC/SCN. 1992. Second report on the world nutrition situation: a report compiled from information available to the ACC/SCN. Administrative Committee on Coordination, Subcommittee on Nutrition. United Nations Subcommittee on Nutrition, Geneva, Switzerland; International Food Policy Research Institute, Washington, DC.

Abbott, P., Konandreas, G., and Benirschka, J. 1993. A model for assessing food security policy alternatives. *In* P. Berck and D. Bigman [*editors*]. Food security and food inventories in developing countries. CAB International, Oxon, U.K.

Adolph, E.F. 1947. Physiology of man in the desert. Interscience Publ., New York, NY. xiii, 357p.

Adams, J.M. 1977. Tropical Science, 19, 1 ODNRI, London.

AFRC. 1992. Food quality and measurement. AFRC Inst Fd Res, U.K.

Agriculture Canada. 1989. Growing together. Ag Can Pubn 5269/E.

Agriculture Canada. 1993. Consumer Food Trends in the 1990s. Agriculture Canada, Ottawa.

AIC. 1986. In search of soil conservation strategies in Canada. Agricultural Institute of Canada, Ottawa. 10 p.

Altieri, M.A. 1983. Agro-ecology: The scientific basis of alternative agriculture. Westview Press, Boulder, CO.

Ames, B.N. 1983. Dietary carcinogens and anticarcinogens. Science 221(4617): 1256-1264.

Ames, B.N., and Gold, L.S. 1990. Environmental pollution and cancer: Some misconceptions. Science and the Law 25 July, 1-33.

APDC. 1991. Biotechnology for Asian agriculture: public policy implications. Edited by I.P. Getubig, Jr., V.L. Chopra, and M.S. Swaminathan. Asian and Pacific Development Centre, Kuala Lumpur, Malaysia. 324p.

Baker, F.W.G. [*editor*]. 1989. Drought resistance in cereals. Published for ICSU Press by CAB International, Wallingford, Oxon, U.K. viii, 222p.

Baker, F.W.G., and Terry, P.J. [*editors*]. 1991. Tropical grassy weeds. Published for CASAFA by CAB International, Wallingford, Oxon, U.K. viii, 203p.

Baum, W.C. 1986. Partners against hunger: the Consultative Group on International Agricultural Research. World Bank, Washington, DC. xiii, 337p.

Bayliss, B.T. 1992. Transport policy and planning: an integrated analytical approach. World Bank, Washington, DC. ix, 68p.

Beaton, G.H. 1991. Human nutrient requirement estimates. Food, Nutrition and Agriculture. Vol. 1, 3-15. FAO, Rome.

Bender, A.E. 1987. Eating to live or living to eat. Plenary Lecture: IUFOST Congress VII, Singapore.

Benedict, F.G. 1917. A study of prolonged fasting. Carnegie Institution of Washington, Washington, DC. 416p.

Benedict, F.G., Miles, W.R., Roth, P., and Smith, H.M. 1919. Human vitality and efficiency under prolonged restricted diet. Carnegie Institution of Washington, Washington, DC.

Berck, P., and Bigman, D. 1993. Food security and food inventories in developing countries. CAB International, Oxon, U.K. x, 381p.

Bigman, D. 1993. The measurement of food insecurity. *In* P. Berck and D. Bigman [*editors*]. Food security and food inventories in developing countries. CAB International, Oxon, U.K.

Block, G., Dressor, C.M., Hartman, A.M., and Carroll, M.D. 1985. Nutrient sources in the American diet. Amer. J. Epidemiol. 122: 27-40.

BMA. 1950. Report of the Committee on Nutrition: British Medical Association. London, U.K.

Boeringa, R. 1980. Alternative methods of agriculture. Elsevier Scientific Publ. Co., Amsterdam and New York. vi, 199p.

Bo-Linn, G.W. 1982. Starch blockers - their effects on calorie absorption from a high starch meal. N. Engl. J. Med. 307: 1413-1418.

Bourne, M.C. 1977. Post-harvest food losses. Cornell University Press, Ithaca, NY.

Brown, L.R. 1987. Sustaining world agriculture. Worldwatch Institute, Washington, DC.

Brown, L.R., and Kane, H. 1993. Full House: Reassessing the Earth's Population Carrying Capacity. Worldwatch Environmental Alert Series.

Brown, L.R., and Wolf. 1984. Soil erosion. Worldwatch Pap. No. 60. Worldwatch Institute, Washington, DC.

Brown, M., and Goldin, I. 1992. The future of agriculture: Developing country implications. OECD, Paris.

Brubacher, G., Muller-Mulot, W., and Southgate, D.A.T. 1985. Methods for the determination of vitamins in food recommended by COST 91. Elsevier Appl. Sci. Pub., Barking, Essex, U.K. x, 166p.

Bunting, A.H. 1987. Agricultural environments: characterization, classification and mapping. CAB International, Oxon, U.K. 335p.

Busch, L., and Lacy, W.B. 1984. Food security in the United States. Westview Press, Boulder, CO. xxii, 430p.

Butler, L. 1994. Role of polyphenols in sustainable production and utilization of sorghum and millet. INTSORMIL (Univ. of Nebraska) Annu. Rep. 1994: 13-17.

Buzina, R. 1992. Recent evolution of food habits in Mediterranean populations, pp. 67-74. *In* P. Leathwood, M. Horisberger, and W.P.T. James [*editors*]. For a better nutrition in the 21st Century. Raven Press, NY. xiv, 258p.

CABI. 1991. The biodiversity of microorganisms and invertebrates. *In* D.L. Hawksworth [*editor*]. Workshop on the ecological foundations of sustainable agriculture (WEFSA 1), London, 26-27 July 1990. CAB International, Oxon, U.K. xx, 302p.

CAEFMS. 1990. Economic dimensions of sustainable agriculture. Canadian Agricultural Economics and Farm Management Society. Submission to the Science Council of Canada and the Agricultural Institute of Canada. CAEFMS, Ontario. v, 45p.

Cantrell, R.P. 1990. Quality protein maize. *In* Sorghum nutritional quality. Purdue University, IL, USA.

Carroll, K.K. 1986. Biological effects of fish oils in relation to chronic diseases. Lipids

21: 731-734.

Carroll, W.R. 1958. Management and the food technologist. Can. Food Ind. 29: 7-8.

Carruthers, I. 1993. Going, Going, Gone! Tropical agriculture as we knew it. Trop. Agric. Assoc. Newsl. UK 13(3): 1-5.

Carruthers, I., and Morrison, J. 1994. 2020 Vision - Dramatic changes in the world agricultural and industrial production systems. IIMI (Int. Irrigation Manage.Inst., Colombo)Review 8(1): 14-20.

CASAFA. 1991. Sustainable agriculture and food security. Rep. to UN Conf. on Environmment and Development, Rio by ICSU Committee on the Appl. of Sci to Agriculture, Forestry and Aquaculture. ICSU Press, Paris.

CDH Australia. 1983. Dietary Guidelines for Australians. Commonwealth Department of Health, Canberra, Australia.

CGIAR. 1972-1994. Annual Reports of the Consultative Group on International Agricultural Research. World Bank, Washington, DC.

CGIAR. 1995. Renewal of the CGIAR: Sustainable agriculture for food security in developing countries. CGIAR, World Bank, Washington, DC.

CGIAR/TAC. 1989. Sustainable agricultural production: implications for international agricultural research. Technical Advisory Committee to the Consultative Group on International Agricultural Research. FAO Tech. Pap., TAC Sect., FAO, Rome. xxiii, 131p.

Chandler, R.F. 1982. An adventure in applied science: a history of the International Rice Research Institute. Int. Rice Res. Inst., Los Bacnos, Philippines. 233p.

Chapman, N. 1992. SNE Advocates for Food Assistance and Nutrition Education: A 25-year History. J. Nutr. Educ. 24[1]: 11S.

Check, W. 1983. Eat, drink and be merry. J. Amer. Med. Assoc. 250(6): 701-711.

Chisnall, P.M. 1986. Marketing research. 3rd ed., McGraw-Hill, London, U.K. x, 352p.

CIMMYT. 1992. Enduring designs for change: an account of CIMMYT's research, its impact, and its future directions. International Maize and Wheat Improvement Center, Mexico. 118p.

CIMMYT. 1995. Mimicking Mother Nature to increase wheat's genetic diversity. Res. Briefs, Feb. 1995, CIMMYT, Mexico.

CIP. 1984. Potatoes for the developing world. International Potato Center, Lima, Peru.

Clancy, K.L., and Bowering, J. 1992. Need for emergency food: poverty problems and policy responses. J. Nutr. Educ. 24(1): 12S-17S.

Cohen, B.E., Chapman, N., and Burt, M.R. 1992. Food Sources and Intake of Homeless Persons. J. Nutr. Educ. 24(1): 45S-51S.

Collinson, M.P., and Tollens, E. 1994. The impact of the International Agricultural Research Centers. CGIAR, World Bank, Washington, DC.

Coultate, T.P. 1984. Food, the chemistry of its components. Royal Society of Chemistry, London, U.K. 197p.

Creightney, C.D. 1993. Transport and economic performance: a survey of developing countries. Tech. Pap. 232, World Bank, Washington, DC. ix, 40p.

CRSP/USAID. 1993. Global research for sustainable development. CRSP Council, USAID, Washington, DC.

Curtis, R.F. 1986. Food science, food quality and food standards. Proc. Inst. Food Sci. Tech. (U.K) 19(4): 151-164.

De Boer, A.J., Yazman, J.A., and Raun, N.S. 1994. Animal agriculture in developing countries: Technology dimensions. Winrock Int. Morrilton, Arkansas, USA.

Dekker, J. 1991. Pesticides. *In* Sustainable agriculture and food security. Rep. to UN Conf. on Environmment and Development, Rio by ICSU Committee on the Appl. of Sci to Agriculture, Forestry and Aquaculture. ICSU Press, Paris.

Delgado, C., and Mellor, J. 1987. A structural view of policy issues in African agricultural development. Amer. J. Agric. Econ. 69: 389-391.

De Soto, H. 1993. The missing ingredient. *In* The Future Surveyed: 150 Economist Years, pp. 8-12. Economist 11: Sept. 1993.

DND Canada. 1994. National Defence, Canada. Private communication.

Donaldson, E.M. 1986. Controlled reproduction techniques in Pacific salmonid culture. Fish. Physiol. Biochem. 2: 9-24.

Donaldson, E.M. 1988. Science and the future of aquaculture, p. 299-309. *In* Proc. of the Int. Aquaculture Congress. British Columbia Pavilion Corporation, Vancouver, BC.

Doyle, M.P. 1994. The emergence of new agents of foodborne disease in the 1980s. Food Res. Int. 27(3): 219-226.

Economist. 1994. Infrastructure in Asia. 26 Feb. p. 66; Electricity in China. 19 Mar. pp. 81-82; Inequality: for richer, for poorer. 5 Nov. pp. 19-21.

Economist. 1995a. Out of Africa. 10 June p. 72, 1 July p. 8.

Economist. 1995b. Springtime for scientists in Georgia. 15 Feb. pp. 83-84.

Edwards, C.A., Lal, R., Madden, P., Miller, R., and House, G. 1990. Sustainable agricultural systems. Soil and Water Conservation Society, Ankeny, IA, USA. xvi, 696p.

El-Lakany, M.H. 1991. Sustainable agricultural development under marginal conditions: An example from Egypt. *In* Sustainable agriculture and food security. Rep. to UN Conf. on Environmment and Development, Rio by ICSU Committee on the Appl. of Sci to Agriculture, Forestry and Aquaculture. ICSU Press, Paris.

ERG. 1986. Energy research: directions and issues for developing countries. Energy Research Group. IDRC 250e. IDRC, Ottawa. ix, 184p.

FAO. 1977. Post-harvest food losses. Rep. AGPP, Misc/27.

FAO. 1984. Land, food and people. FAO, Rome.

FAO. 1992a. The state of food and agriculture 1992. FAO, Rome.

FAO. 1992b. Production year book 1992. FAO, Rome.

FAO. 1993. World agriculture: towards 2010. N. Alexandratos [*editor*]. FAO, Rome.

FAO/WHO. 1962. Calcium requirements. Report of FAO/WHO Expert Group Calcium requirements. WHO Tech. Rep. Ser. No. 230: 54p.

FAO/WHO. 1965a. Protein requirements. Report of FAO/WHO Expert Group on Protein Requirements. WHO Tech. Rep. Ser. No. 301: 71p.

FAO/WHO. 1965b. Requirements of vitamin A, thiamine, riboflavin, and niacin. Report of FAO/WHO Expert Group. WHO Tech. Rep. Ser. No. 362: 86p.

FAO/WHO. 1973. Energy and protein requirements. Report of FAO/WHO Ad Hoc Expert Committee. WHO Tech. Rep. Ser. No. 522: 118p.

FAO/WHO. 1988. Requirements of vitamin A, iron, folate and vitamin B12. Report of

a joint FAO/WHO Expert Consultation. FAO Food Nutr. Ser. No. 23: 107p.
FAO/WHO. 1992. Plan of action for nutrition. FAO Publication ICN/92/2-Rev 2 Dec 1992.
FNB/NRC. 1989. Recommended Dietary Allowances. 10th ed., Food and Nutrition Board, Commission on Life Sciences, National Research Council (U.S.); National Academy Press, Washington, DC. x, 284 p.
FRAC. 1991. Hunger survey index. Food Research and Action Center, Washington, DC.
Fennema, O. 1985. Chemical changes in food during processing. AVI Pub. Co., Westport, CN, USA.
Fernstrom, J.D., and Wurtman, R.J. 1971, 1972. Brain serotonin content. Science 174: 1023-1026; Science 178: 414-417.
Food Tech. 1987. Rapid methods for microbiological analysis of food. Food Tech. July 56-71.
Food Tech. 1993. America's food research needs: Into the 21st Century. Special report. Food Tech. 47(3): 2S-39S.
Freudenberger, C.D. 1984. International trade and American food security. *In* L. Busch and W.B. Lacy [*editors*]. Food security in the United States. Westview Press, Boulder, CO. xxii, 430p.
Garnick, R.L., Sulli, N.J., and Papa, P.A. 1988. The role of quality control in biotechnology: An analytical perspective. Anal. Chem. 60: 2546-2557.
GASGA. 1978. Priorities for action in grain post-harvest loss. Kansas State Univ, Manhattan, KS.
Gamble, J.L. 1944. Water requirements of castaways. Proc. Amer. Phil. Soc. 88: 151-158.
Garrow, J.S. 1983. Starch blockers are ineffective in man. Lancet 1: 60-65.
Gianessi, L. 1993. Chemical-free farming. Issues in Science & Technology 10(1): 29-36.
Glomset, J.A. 1985. Fish, fatty acids and human health. N. Engl. J. Med. 312(19): 1253-1254.
Golitsyn. G.S., McBean, G.A., and Sanhueza, E. 1992. Atmosphere and Climate. *In* An agenda of science for environment and development into the 21st Century. ICSU/Cambridge Univ. Press, U.K.
Greathead, D.J. 1987. Prospects for biological control in the tropics. Insect Sci. Appl. 8: 787-789.
Greathead, D.J. 1989. Prospects for the use of natural enemies in combinations with pesticides. Tech. Bull. #117, Food and Fertilizer Tech. Centre, Tapei.
Greathead, D.J. and Waage, J.K. 1983. Opportunities for biological control of agricultural pests in developing countries. World Bank Tech. Pap. 11, World Bank, Washington, DC. 44p.
Greenland, D.L., and Lal, R. 1977. Soil conservation and management in the humid tropics. J. Wiley, New York, NY. 283p.
H&W Canada. 1988. Canadian Guidelines for Healthy Weights. Report of an Expert Group convened by Health Services and Promotion Branch, Health and Welfare Canada, Ottawa. 126p.
H&W Canada. 1989. Approaches to assessing the safety of crops developed through wide-cross hybridisation techniques. Food Directorate, Health and Welfare Canada, Ottawa.

H&W Canada. 1990. Nutrition Recommendations: Report of the Scientific Review Committee. Health and Welfare Canada, Ottawa. 208p.

H&W Canada. 1992. Canada's Food Guide to Healthy Eating. Health and Welfare Canada, Ottawa.

Hall, D.W. 1970. Storage of food grains. FAO Agricult. Dev. Pap. 90: 20.

Harris, R.S., and Karmas, E. 1975. Nutritional evaluation of food processing. AVI Pub. Co., Westport, CN, USA. xiii, 670p.

Harris, K.L, and Lindblad, C.J. 1978. Post-harvest loss assessment. Amer. Assoc. Cereal Chem., Minneapolis.

Harrison, K., Henley, D., Riley, H., and Shaffer, J. 1987. Improving marketing systems in developing countries: Experiences from Latin America. MSU Development Papers Res. Rep. 8, Mich. State Univ.

Hayner, J.H., and Proctor, B.E. 1953. Possible use of atomic fission products for food sterilization. Food Technol. 7(1): 6-10.

Hegsted, D.M. 1975. Dietary standards. J. Amer. Diet. Assoc. 66: 13-21.

Herold, P.M., and Kinsella, J.E. 1986. Fish oil consumption and decreased risk of cardiovascular disease. Amer. J. Clin. Nutr. 43: 566-598.

Hine, J.L., Riverson, J.D.N., and Kwakye, E.A. 1983. Accessibility, transport, and food marketing in the Ashanti region of Ghana. Transport and Road Res Lab, Berkshire, U.K.

Hollingsworth, P. 1994. The perils of product development. Food Technol. 48(6): 80-88.

Hollingsworth, P. 1995. Food research: Cooperation is the key. Food Technol. 49(2): 67-74.

Hulse, J.H. 1981. Human implications of protein utilization, p.1-17. *In* E.D. Murray, D.H. Lees, and D.W. Stanley [*editors*]. Utilization of World Protein Resources. Food and Nutrition Press, Westport, CN.

Hulse, J.H. 1991. Nature, composition and utilization of grain legumes, p. 11-27. *In* Uses of Tropical Grain Legumes. Pub. ICRISAT, Patencheru, India.

Hulse, J.H. 1991a. Global perspectives on sustainable development: Implications for agriculture. Can. J. Agric. Econ. 39: 541-551.

Hulse, J.H. 1992. True knowledge is itself power. Food Res. Int. 25: 237-245.

Hulse, J.H. 1993. Agriculture, food and the environment. Food Res. Int. 26: 455-469.

Hulse, J.H. 1994a. Nature, composition and utilization of food legumes. *In* F.J. Muehlbauer and W.J. Kaiser [*editors*]. Expanding the production and use of cool season food legumes. Kluwer Acad. Publishers, Dordrecht, Netherlands.

Hulse, J.H. 1994b. Renaissance: Science for prosperity and progress. Food Res. Int. 27: 83-96.

Hulse, J.H., and Laing, E.M. 1974. Nutritive value of triticale protein and the proteins of wheat and rye. Int. Devel. Res. Centre, Ottawa. 183p.

Hulse, J.H., and Pearson, O. 1980. Nutritional status of rural population in the Sahel, pp. 86-92. *In* Nutritional status of the population of the semi-arid tropics. Int. Devel. Res. Centre, Ottawa.

Hulse, J.H., Laing, E.M., and Pearson, O.E. 1980. Sorghum and the Millets: their composition and nutritive value. Academic Press, London. xvii, 997p.

Hulse, J.H., and Escott, V.J. 1986. Drought-Inevitable and Unpredictable. Interdiscip. Sci.

Rev. 11(4): 346-358.
Huysmanns, A.A.C. 1981. Post-harvest losses, p. 1099-1126. *In* J.M. Manassah and E.J. Briskey [*editors*] Advances in food-processing systems for semi-arid lands. Academic Press, New York, NY.
Huxley, P.A. 1983. Plant research and agroforestry. International Council for Research in Agroforestry [ICRAF] Nairobi, Kenya. xv, 617p.
IARI. 1993. Neem in agriculture. Indian Agric. Res. Inst., New Delhi, India.
IBPGR. 1991. International Board for Plant Genetic Resources: Annual Report 1991. IBPGR/IPGRI, Rome.
ICRAF. 1991. Agroforestry systems. ICRAF/Kluwer Acad Pub. 15(2-3): 91-308.
ICRISAT. 1995. Sustainable natural resource management: Options to arrest land degradation in desert margins of Sub-Saharan Africa. ICRISAT, Andhra Pradash, India.
ICSU. 1992. An agenda of science for environment and development into the 21st Century. International Council Sci. Unions, Cambridge University Press, New York, NY.
ICSU/WMO. 1990. Global climate change. World Climate Res Programme: ICSU, Paris.
IDRC. 1980. Nutritional status of the population of the semi-arid tropics. Int. Devel. Res. Centre, Ottawa.
IDRC 1981. A decade of learning. IDRC 170e. Int. Dev. Res. Centre, Ottawa.
IDRC. 1983. The very basis of genius. Int. Devel. Res. Centre, Ottawa.
IDRC. 1986. Biotechnology: Opportunities and constraints. IDRC-MR1-10er Int. Devel. Res. Centre, Ottawa. 162p.
IFBC. 1990. Biotechnologies and food: Assuring the safety of foods produced by genetic modification. Regulat. Toxicol. Pharmacol. 12(3): S1-S196.
IFST. 1993. Keynote: Newsletter of UK Inst. of Food Sci. & Tech. Dec. 1993.
IFPRI. 1995a. Population and food in the Twenty-First Century. Int. Food Policy Res. Inst., Washington, DC.
IFPRI. 1995b. A 2020 vision for food, agriculture and environment: Recommended action. Draft publication. IFPRI, Washington, DC.
IFPRI/RF. 1992. International agricultural research: Facts and figures. Int. Food Policy Res. Inst., Washington, DC.
IGBP/ICSU. 1992. Global change: Reducing uncertainties. *In* Global Biosphere Programme, Stockhom, Sweden.
IIASA. 1990. Global soil change. Report of IIASA, ISSS, UNEP Task Force on the Role of Soil in Global Change. Int. Inst. Appl. Syst. Anal., Laxenberg, Austria. 110p.
ILRAD. 1990, 1991. Annual Reports. International Laboratory for Animal Diseases. Nairobi, Kenya.
IPGRI. 1993. Diversity for development. Int. Plant Genet. Res. Inst., Rome.
IRRI. 1984. Genetic manipulation of crops. Int. Rice Res. Inst., Los Banos, Philippines.
IRRI. 1985. Biotechnology in international agricultural research. Int. Rice Res. Inst., Los Banos, Philippines. 435p.
IRRI. 1993. IRRI rice almanac. Int. Rice Res. Inst., Philippines.
IRRI. 1994. IRRI redesigns rice plant to yield more grain. IRRI Reporter 4/94: 1-2. Int. Rice Res. Inst. Los Banos, Philippines.

ISRIC. 1990. Global assessment of soil degradation. Int. Soil Ref. Info. Centre, Wageninigen, Netherlands.

IUNS/CABI. 1983. Recommended dietary intakes around the world. Nutr. Abstr. Rev. 53(11): 939-1117.

Jain, H.K. 1983. Agriculture and environment. Indian Farming, June 1983, pp. 5-19.

Janvry, A. de, and Sadoulet, E. 1993. New ideas for poverty alleviation in the 1990s. *In* P. Berck and D. Bigman [*editors*]. Food security and food inventories in developing countries. CAB International, Oxon, U.K. x, 381p.

Jones. S.A. 1995. Fat replacers: The Broad Perspective. The World of Ingredients (May-June): 8-13.

Juliano, B.O. [*editor*]. 1985. Rice: chemistry and technology. Amer. Assoc. Cereal Chem., St. Paul, MN, USA. 774p.

Kahn, H. 1982. The coming boom: economic, political, and social. Simon & Schuster, NY. 237p.

Keys, A. 1957. Diet and the epidemiology of coronary heart disease. J. Amer. Med. Assoc. 164: 1912-1919.

Keys, A. 1975. Coronary heart disease: the global picture. Atherosclerosis 22: 149-192.

Keys, A. 1980. Alphalipoprotein [HDL] cholesterol in the serum and risk of coronary heart disease and death. Lancet 2: 603-606.

Keys, A. 1984. Serum cholesterol response to dietary cholesterol. Amer. J. Clin. Nutr. 40: 351-359.

Keys, A., Brozed, J., Henschel, A., Mickelson, O., and Taylor, H.L. 1950. The biology of human starvation. Univ. of Minnesota Press, Minnesota.

Keys, A., Menotti, A., and Karvonen, M.J. 1980. Diet and a 15-year death rate in a seven countries study. Amer. J. Epidemiol. 124: 903-915.

Koppel, B. 1984. International trade and American food security. *In* L. Busch and W.B. Lacy [*editors*]. Food security in the United States. Westview Press, Boulder, CO. xxii, 430p.

Kromhout, D., Bosschieter, E.B., and Coulander, C. de L. 1985. Inverse relation between fish consumption and 20 year mortality from coronary heart disease. N. Engl. J. Med. 312(19): 1205-1209.

Lacy, W.B., and Busch, L. 1984. The role of agricultural research for US food security. *In* L. Busch and W.B. Lacy [*editors*]. Food security in the United States. Westview Press, Boulder, CO. xxii, 430p.

Lal, R. 1989. Conservative tillage for sustainable agriculture. Adv. Agron. 42: 85-197.

Lancet. 1979. Editorial. Lancet 1: 249-250.

Landegren, U., Kaiser, R., and Hood, L. 1988. DNA diagnostics: Molecular techniques and automation. Science 242: 229-237.

Lane, H.W., and Rice, B.L. 1991. Nutritional requirements for space station crews. NASA Conf. Publ. 3146, Houston, TX.

Lane, H.W., Pool, S.L., and Huntoon, C.L. 1992. Nutritional requirements for extended duration Orbiter missions. [30-90d] NASA Publ. JSC 32283.

Lane, H.W., and Schulz, L.O. 1992. Nutritional questions relevant to space flight. Ann. Rev. Nutr. 12: 257-278.

Lane, H.W., Le Blanc, A.D., Potch, A.L., and Whitson, P.A. 1993. Nutrition and human physiological adaptations to space flight. Amer. J. Clin. Nutr. 58: 583-588.

League of Nations. 1936. The problem on nutrition. Vol 1 & 2 Technical Committee on Nutrition, Geneva, Switzerland.

League of Nations. 1937. The relation of nutrition to health, agriculture and economic policy. League of Nations, Geneva, Switzerland.

LeVeen, E.P. 1984. Domestic food security and increasing competition for water. *In* L. Busch and W.B. Lacy [*editors*]. Food security in the United States. Westview Press, Boulder, CO. xxii, 430p.

LSRO. 1987. Physiological effects and health consequences of dietary fibre. Life Sciences Research Office of the Federation of American Societies for Experimental Biology. Bethesda, MD.

Lutz, W., and Prinz, C. 1991. Scenarios for world population in the next century. WP-91-92. International Institute for Applied Systems Analysis, Vienna.

Madden, J.M. 1994. The enterics as foodborne pathogens. Food Res. Int. 27: 227-232.

Mann, D.W. 1981. Fewer people for a better world: A plea for negative population growth. Environ. Conserv. 8(4): 260-261.

Matsumoto, M. 1989. Recent trends in domestic food programs. Nat. Food Rev. 12(4): 34-36.

McBean, G.A., Golitsyn, G.S., Sanhueza, E. 1992. Atmosphere and Climate. *In* An agenda of science for environment and development into the 21st Century. International Council Sci. Unions, Cambridge University Press, New York, NY.

McCalla, A.F. 1994. Agriculture and food needs to 2025: Why we should be concerned. Sir John Crawford Memorial Lecture, CGIAR, World Bank, Washington, DC.

McCance, R.A., and Widdowson, E.M. 1960. The Composition of Foods UK. Med Res Council, H.M.S.O., London. xii, 417p.

Melville, R. 1966. Continental drift: Mesozoic continents and the migrations of the angiosperms. Nature 211: 116-120.

Mertens, V., and Hlynka, I. 1969. Protein content of Canadian wheat: 1927-1968. Board of Grain Commissioners, Winnipeg, MB, Canada.

Mehta, K. 1995. New policies and their effect on food processing in Europe. Food Technol. Int. Europe 1995: 23-24.

Miller, G.T., and Miller, N.E. 1975. Plasma-high-density-lipoprotein concentration and development of ischaemic heart disease. Lancet 1: 16-19.

Miller, N.E., Thelle, D.S., Forde, O.H., and Mjos, O.D. 1977. The Tromso Heart Study. Lancet 1: 965-968.

Mintz, S.W. 1989. Food and culture. *In* P.M. Hirschoff and N.G. Koptler [*editors*]. Completing the food chain: strategies for combatting hunger and malnutrition. Smithsonian Institution Press, Washington, DC. xxii, 193p.

Mitchell, D.O., and Ingco, M.D. 1993. The World Food Outlook. International Economics Dept. World Bank, Washington, DC.

Morris, T.N. 1951. Principles of Fruit Preservation. Chapman & Hall.

Morris, P.M., Neuhauser, L., and Campbell, C. 1992. Food Security in Rural America. J. Nutr. Educ. 24(1): 52S-57S.

Mossel, D. 1989. Food safety and the need for public assurance. Food Sci. Tech. Today [UKIFST] 3(1): 2-4.

MSSRF. 1991. Biotechnology: Reaching the unreached. M.S. Swaminathan Res. Foundation, Madras, India.

Mulongoy, K., Gueye, M., and Spencer, D.S.C. 1990. Biological nitrogen fixation in tropical agriculture. J. Wiley, NY.

NABC. 1989. Biotechnology and sustainable agriculture. Nat. Agric. Biotech. Counc. Rep. Boyce-Thompson Inst. Ithaca, NY, USA.

NAS. 1978. Post-harvest food losses in developing countries. Nat. Acad. Press, Washington, DC.

Nestle, M., and Guttmacher, S. 1992. Hunger in the USA: Rationale, Methods and Policy Implications of State Hunger Surveys. J. Nutr. Educ. 24(1): 18S-22S.

NIH. 1985. National Institutes of Health Consensus Development Conference: Health implications of obesity. Ann. Int. Med. 103(No. 6 Pt. 2).

Norse, D., James, C., Skinner, B.J., and Zhao, Q. 1992. Agriculture, land use and degradation. *In* An agenda of science for environment and development into the 21st Century. International Council Sci. Unions, Cambridge University Press, New York, NY.

NRC/NAS. 1977. NRC World food and nutrition study: the potential contributions of research. U.S. Nat. Acad. Sci., Washington, DC.

NRC/NAS. 1989. Alternative agriculture. U.S. NRC. National Academy Press, Washington, DC.

NRC. 1989. Diet and Health - Implications for reducing chronic disease risk. Committee on Diet and Health, U.S. NRC, National Academy Press, Washington, DC.

NRC/RDAs. 1989. Recommended Dietary Allowances. 10th ed., Food and Nutrition Board, U.S. NRC, National Academy Press, Washington, DC.

O'Hare, W., Mann, T., Porter, K., and Greenstein, R. 1990. Real life poverty in America. Where the American public would set the poverty line. Center on Budget and Policy Priorities. Washington, DC.

Palaniswami, S. 1993. Design of quality control methods for grains in developing countries: A simulation analysis. *In* P. Berck and D. Bigman [*editors*]. Food security and food inventories in developing countries. CAB International, Oxon, U.K. x, 381p.

Pardey, P., Roseboom, J., and Beintema, N. 1994. Agricultural research in Africa: Three decades of development. ISNAR Briefing Paper. International Service to National Agricultural Research, The Hague, Netherlands.

Parikh, J.K. 1986. Energy in post-harvest food systems in south Asia. Energy Policy 14(4): 363-372.

Parker, S.L. 1992. A National Survey of Nutritional Risk among the Elderly [1992]. J. Nutr. Educ. 24(1): 23S.

Parmal, B.S., and Devakumar, C. 1993. Botanical and biopesticides. Westville Pub House, New Delhi, India.

Parry, M.L., Porter, J.H., and Carter, T.R. 1990. Climatic change and its implications for agriculture. Outlook on Agric. 19(1): 9-15.

Patel, P. 1994. DNA fingerprinting in food analysis. Food Tech. Int. Europe 1994, pp. 171-174.

Pearson, D.J., Rix, K.J.B., and Bentley, S.J. 1983. Food allergy: How much in the mind. Lancet 1: 1259-1261.

Pellett, P.L., and Young, V.R. 1980. Nutritional Evaluation of Protein Foods. United Nations University, Tokyo.

Peterson, W. 1993. Agricultural prices: A Third World dilemma. *In* P. Berck and D. Bigman [*editors*]. Food security and food inventories in developing countries. CAB International, Oxon, U.K. x, 381p.

Phillipson, B.E., Rothrock, D.W., and Connor, W.E. 1985. Reduction of plasma lipids, lipoproteins and apoproteins by dietary fish oils in patients with hypertriglyceridemia. N. Engl. J. Med. 312(19): 1210-1254.

Physician Task Force on Hunger in America. 1985. Hunger in America: The growing epidemic. Harvard University School of Public Health. Wesleyan University Press, Middletown, CN. xxii, 231p.

Philogene, B.J.R. 1981. Successes and future prospects for host plant resistance in integrated control systems. *In* The role of biological control in pest management. IOBC/IUBS, Paris.

Philogene, B.J.R. 1988. Utilisation et reglement des pesticides. *Dans* le Tiers Monde Revue Canadienne d'etudes du developpement, pp. 275-278.

Pimental, D. 1984. Energy inputs and food security. *In* L. Busch and W.B. Lacy [*editors*]. Food security in the United States. Westview Press, Boulder, CO. xxii, 430p.

Pimental, D., and Burgess, M. 1980. Energy inputs in corn production. *In* D. Pimental [*editor*]. Handbook of energy utilization in agriculture. CRC Press, Boca Raton, FL, USA.

Pimental, D., and Pimental, M. 1979. Food, energy and society. E. Arnold [*editor*]. London. viii, 165p.

Pimental, D., Oltenacu, P.A., and Nesheim, M.C. 1980. Grass-fed livestock potential: Energy and land constraints. Science 207: 843-848.

Pinstrup-Andersen, P. 1994. World food trends and future food security. Food Policy Report. Int. Food Policy Res, Inst., Washington, DC.

Plucknett, D.L., Smith, N.J.H., and Williams, J.T. 1987. Gene banks and the world's food. Princeton University Press, Princeton, NJ, USA. xv, 247p.

Poppendieck, J.E. 1992. Hunger and Public Policy: Lessons from the Great Depression. J. Nutr. Educ. 24(1): 6S-10S.

Prasad, R. 1986. Fertilizer nitrogen requirements and management. *In* M.S. Swaminathan and S.K. Sinha [*editors*]. Global aspects of food production. Tycooly International Press, Oxford, U.K.

Prior, C. 1989. Biological pesticides for low external input agriculture. Biocontrol News and Info. CAB Internatl

Reganold, J.P., and Papendick, R.I. 1990. Sustainable agriculture. Sci. Amer. 262(2): 112-120.

Reutlinger, S. 1987. Food security, trade and aid. *In* J.P. Gittinger, J. Leslie, and C. Hoisington [*editors*]. Food Policy. John Hopkins University Press, Baltimore, MD.

Roberts, T. 1989. Human illness costs of foodborne bacteria. Amer. J. Agricult. Econ. 71: 468-474.

Robins, G.N. 1890. The fasting man. Br. Med. J. 1, 764, 996, 1444-1446.

Rodale, R. 1983. Breaking new ground: Search for sustainable agriculture. Futurist 1: 15-20.

Rockefeller. 1982. Science for agriculture: Critical issues in American agricultural research. Rockefeller Foundation, NY.

Rosenthal, S.R. 1934. Studies in atherosclerosis: chemical, experimental and morphological. Arch. Pathol. 18, I to V, pp. 473-506, 660-698, 827-842.

Royal College of Physicians. 1984. J. Roy. College of Physicians, Vol. 18.

Sampson, N. 1984. Agricultural land: Basis for food security. *In* L. Busch and W.B. Lacy [*editors*]. Food security in the United States. Westview Press, Boulder, CO. xxii, 430p.

Santorum, A., and Gray, J. 1993. A general approach to the assessment of public stocks. *In* P. Berck and D. Bigman [*editors*]. Food security and food inventories in developing countries. CAB International, Oxon, U.K. x, 381p.

Scrimshaw, N.S. 1976. An analysis of past and present recommended dietary allowances for protein in health and disease. N. Engl. J. Med. 294: 136-142, 198-203.

Scrimshaw, N.S. 1993. The challenge of global malnutrition to the food industry. Food Tech. 47(2): 60-71.

Scriven, F.M., Squires, P., and Watson, A.J. 1988. Consumer evaluation of product concepts. Food Tech. in Australia 40(6): 219-222.

Select Committee on Hunger. 1990. Food Security in the United States Committee Report. US House of Representatives.

Selvedran, R.R. 1984. The plant cell wall as a source of dietary fibre: chemistry and structure. Amer. J. Clin. Nutr. 39: 320-327.

Sen, A. 1981. Poverty and famines: An essay on entitlement and deprivation. Clarendon Press, Oxford, U.K.

Sen, A. 1987. Poverty and entitlements. *In* J.P. Gittinger, J. Leslie, and C. Hoisington [*editors*]. Food Policy. John Hopkins University Press, Baltimore, MD.

Senate of Canada. 1984. Soil at risk Rpt on soil conservation by a Standing Committee of the Senate of Canada. Ottawa.

Shaw, G.B. 1911/1946. Preface to The Doctor's Dilemma, p. 87. Penguin Books, Harmondsworth, U.K.

Shewry, P.R., Halford, N.G., and Tatham, A.S. 1992. High molecular weight subunits of wheat glutenin. J. Cereal Chem. 15: 105-120.

Smith, P.K., and Hoerr, S.L. 1992. A Comparison of Food Bank Users in a Population of Low Income Single Mothers. J. Nutr. Educ. 24(1): 59S-66S.

Solkoff, J. 1985. The politics of food. Sierra Club Books, San Francisco, CA.

Southgate, D.A.T. 1982. Definitions and terminology of dietary fibre. *In* G.V. Vahouny and D. Kritchevsky [*editors*]. Dietary Fibre in Health and Disease. Plenum Press, NY.

Southgate, D.A.T., and J.V.G.A. Durnin. 1970. Calorie conversion factors. Br. J. Nutr. 24: 517-535.

Stephen, A.M. 1994. Increasing complex carbohydrate in the diet. Food Res. Int. 27(1): 69-75.

Stevenson, J.A.F. 1958. Survey of Survival Feeding. Report No. DR 125, Defence Research Board, Dep. of National Defence, Canada.

Stewart, J.W.B., Victoria, R., and Wolman, M.G. 1992. Global cycles. *In* An agenda of science for environment and development into the 21st Century. International Council Sci. Unions, Cambridge University Press, New York, NY.

Stoskopf, N.C. 1993. Plant breeding: Theory and practice. Westview Press, Boulder, CO, USA.

Surgeon-General. 1988. The Surgeon-General's Report on Nutrition and Health. Publ. U.S. Dep. of Health and Human Services, Washington,DC. xxxvii, 727p.

Swaminathan, M.S., and Sinha, S.K. 1986. Global aspects of food production. Tycooly International Press, Oxford, U.K.

Swinkels, R.A., and Scherr, S.J. 1991. Economic analysis of agroforestry systems. ICRAF, Nairobi.

Tabucanon, M.T. 1993. Applications of multiple-criteria decision making methodology in food policy analysis. *In* P. Berck and D. Bigman [*editors*]. Food security and food inventories in developing countries. CAB International, Oxon, U.K. x, 381p.

TAC/CGIAR. 1990. Expansion of the CGIAR system. TAC, FAO, Rome.

TAC/CGIAR. 1994a. CGIAR Medium-term resource allocations 1994-98. CGIAR/TAC Secretariat, FAO, Rome.

TAC/CGIAR. 1994b. Review of CGIAR priorities and strategies. CGIAR/TAC Secretariat, FAO, Rome.

TAC/CGIAR. 1994c. Stripe study of genetic resources in the CGIAR. CGIAR/TAC Secretariat, FAO, Rome.

Tkachuk, R. 1969. Nitrogen to protein conversion factors for cereals and oilseed meals. Cereal Chem. 46: 206-218.

Torun, B., Young, V.R., and Rand, W.M. 1981. Protein-energy requirements of developing countries: Evaluation of new data. United Nations University, Tokyo, Japan. viii, 268p.

Trowell, H. 1972. Crude fibre, dietary fibre and atherosclerosis. Atherosclerosis 16: 138-140.

Trowell, H. 1972a. Dietary fibre and atherosclerosis. Amer. J. Clin. Nutr. 25: 926-930.

Trowell, H., Southgate, D.A.T., Wolever, T.M.S., Leeds. A.R., Gassull, M.A., and D.J.A. 1976. Dietary fibre redefined. Lancet 1: 967.

Trowell, H., Godding, E., Spiller, G., and Briggs, G. 1978. Fibre bibliographies and terminology. Amer. J. Clin. Nutr. 31: 1489-1490.

Truswell, A.S. 1982. Pop diets for weight reduction. Br. Med. J. 285(6354): 1519-1520.

Truswell, A.S. 1985. Food sensitivity. Br. Med. J. 291: 951-955.

Tso, P., and Weidman, S.W. 1985. Absorption and metabolism of lipids in humans, pp. 1-15. *In* M. Horisberger and U. Bracco [*editors*]. Lipids in Modern Nutrition. Raven Press, NY.

UN (United Nations). 1988. United Nations Population Projections. United Nations, NY.

UNEP. 1982. Losses of agricultural land. *In* World Environment United Nations Environment Programme, Nairobi.

UNDP. 1993. Human Development Report 1993. Oxford University Press, NY.

UNU. 1981. Protein-Energy Requirements of Developing Countries. B. Torun, V.R. Young, and W.M. [*editors*]. UN University, Tokyo.

USDA. 1981. A time to choose: Summary Rpt on the structure of agriculture in the USA. USDA, Washington, DC.

USDA. 1984-1990. Nationwide Food Consumption Surveys, U.S. Department of Agriculture, Human Nutrition Information Service, Hyattsville, MD.

USDA. 1986-1988. Agricultural resources: Cropland, water, conservation Situation and outlook. Reports AR-4, AR-8, AR-12 USDA Washington.

Waage, J.K. 1991. Biodiversity as a resource for biological control *In* The biodiversity of microorganisms and invertebrates. *In* D.L. Hawksworth [*editor*]. Workshop on the ecological foundations of sustainable agriculture (WEFSA 1), London, 26-27 July 1990. CAB International, Oxon, U.K. xx, 302p.

Waage, J.K., and Greathead, D.J. 1988. Biological control: challenges and opportunities. Phil. Trans. Roy. Soc., London B 318: 111-128.

WARDA. 1991. Ann Rep. W African Rice Dev Association, Ivory Coast.

WCED. 1987. Food 2000: WCED Panel on Food Security. World Commission on Environment and Development. Zed Books, London. xi, 131p.

Westing, A.H. 1981. A world in balance. Environ. Conserv. 8(3): 177-183.

WFP. 1993. 1993 Food Aid Review. World Food Programme, Rome.

WHO. 1968. Nutritional anaemias. WHO (World Health Organization) Tech. Rep. Ser. No. 405: 37p.

WHO. 1984. The role of food safety in health and development. WHO (World Health Organization)Tech. Rep. Ser. No. 705.

WHO. 1985. Energy and protein requirements. Rep. of Joint FAO/WHO/UNU Expert Consultation. WHO (World Health Organization) Tech. Rep. Ser. 724: 206p.

Williams, M.A.J., and Balling, R.C. 1994. Interactions of desertification and climate. UN Environmental Program, Nairobi.

Williams, P.C. 1974. Errors in protein testing and consequences. Cereal Sci. Today 19: 280-282, 286.

World Bank. 1988. Road deterioration in developing countries: causes and remedies. World Bank Policy Study, Washington, DC.

World Bank. 1992. World development report 1992. World Bank, Washington, DC.

World Bank. 1994. Global economic prospects and the developing countries. World Bank, Washington, DC.

World Bank. 1995. World Development Report 1995. World Bank, Washington, DC.

WRC. 1978. The nations water resources 1975-2000. Water Resources Council, U.S. Government Printing Office, Washington, DC.

WRI. 1992. World resources 1992-93: guide to the global environment. World Resources Inst., Washington, DC.

Yetiv, J.Z. 1987. Popular Nutritional Practices: A Scientific Appraisal. Popular Medicine Press, Toledo, OH, USA. 318p.

Yudelman, M. 1994. Demand and supply of foodstuffs up to 2050 with special reference to irrigation. IIMI (Int. Irrigation Manage. Inst., Colombo) Review. 8(1): 4-14.

ACRONYMS AND ABBREVIATIONS

AAAS	American Association for the Advancement of Science
ACIAR	Australian Centre for International Agricultural Research
ACFD	African Centre for Fertilizer Development
AIC	Agricultural Institute of Canada
AOAC	Association of Official Analytical Chemists
AVRDC	Asian Vegetable Research and Development Centre
BMI	Body Mass Index
CABI	CAB International
CARD	Center for Agricultural and Rural Development (USA)
CASAFA	ICSU Scientific Committee on the Application of Science to Agriculture, Forestry and Aquaculture
CGIAR	Consultative Group on International Agricultural Research
CIAT	International Centre for Tropical Agriculture
CIDA	Canadian International Development Agency
CIFOR	Centre for International Forestry Research
CIMMYT	International Centre for Maize and Wheat Improvement
CIP	International Potato Centre
CIRAD	Centre de Coopération Internationale en Recherche Agronomique pour le Développement
CRSP	Collaborative Research Support Program (USAID)
DAC	Development Assistance Committee of OECD
DMI	Desert Margins Inititative (CGIAR)
EPA	U.S. Environment Protection Agency
ER	Energy Ratio
EU	European Union
FAO	Food and Agriculture Organization of the United Nations
GATT	General Agreement on Tariffs and Trade
GDP	Gross Domestic Product
GER	Gross Energy Requirement
GNP	Gross National Product
IARC	International Agricultural Research Centre
IBRD	International Bank for Reconstruction and Development
IBSRAM	International Board for Soil Research and Management
ICARDA	International Centre for Research in Dry Areas
ICLARM	International Centre for Living Aquatic Resources Management
ICRAF	International Centre for Research in Agroforestry

ICRISAT	International Crops Research Institute for the Semi-Arid Tropics
ICSU	International Council of Scientific Unions
IDA	International Development Association
IDRC	International Development Research Centre
IFAD	International Fund for Agricultural Development
IFBC	International Food Biotechnology Council
IFDC	International Fertilizer Development Centre
IFPRI	International Food Policy Research Institute
IGBP	International Geosphere-Biosphere Programme
IIASA	International Institute for Applied Systems Analysis
IIBC	International Institute of Biological Control
IIMI	International Irrigation Management Institute
IITA	International Institute for Tropical Agriculture
ILCA	International Livestock Centre for Africa
ILRAD	International Laboratory for Research on Animal Diseases
ILRI	International Livestock Research Institute
IMF	International Monetary Fund
INCD	International Convention on Desertification
INIBAP	International Network for the Improvement of Bananas and Plantains
INTSORMIL	Sorghum/Millet Collaborative Research Support Program
IPGRI	International Plant Genetic Resources Institute
IPM	Integrated Pest Management
IRRI	International Rice Research Institute
ISNAR	International Service for National Agricultural Research
ISSS	International Soils Science Society
IUCN	International Union for the Conservation of Nature
IUFOST	International Union of Food Science and Technology
IUNS	International Union of Nutrition Sciences
NARS	National Agricultural Research Systems
NAS	National Academy of Sciences (USA)
NRCCAN	National Research Council of Canada
NRCUS	National Research Council (USA)
NGO	Non-government organization
NRI	Natural Resources Institute (UK)
ODA	1 Official Development Assistance
ODA	2 Overseas Development Agency (UK)
OECD	Organization for Economic Cooperation and Development
RDA	Recommended Daily Allowance of specific nutrients
RDI	Recommended Daily/Dietary Intake of specific nutrients
RNI	Recommended Nutrient Intake
TAC	Technical Advisory Committee of the CGIAR
TGE	Tonnes of Grain Equivalent: An index of food energy production
TOE	Tonnes of Oil Equivalent: An index of fuel energy

TWAS	Third World Academy of Sciences
UN	United Nations
UNCED	United Nations Conference on the Environment and Development
UNDP	United Nations Development Programme
UNEP	United Nations Environmental Programme
UNESCO	United Nations Educational, Scientific and Cultural Organization
UNICEF	United Nations Children's Fund
USAID	United States Agency for International Development
USDA	United States Department of Agriculture
USFDA	United States Food and Drug Administration
WARDA	West African Rice Development Association
WCRP	World Climate Research Programme
WFP	World Food Programme
WHO	World Health Organization
WRC	Water Resources Council (USA)
WRI	World Resources Institute

INDEX